IS THIS
THE AGE OF
FRANKENSTEIN?

Scientists today are embarking on a new venture: a genetic revolution!

Long the realm of fiction, the advent of man creating man is upon us. How will this affect *your* future?

The Brave New World which Huxley described is now on our doorstep. Bruce Anderson examines in-depth the steps which could well lead to the genetic engineering of mankind in our lifetime.

D1532052

Let Us Make Man

Let Us Make Man

by

Bruce L. Anderson

Logos International
Plainfield, New Jersey

LET US MAKE MAN
Copyright © 1980 by Bruce L. Anderson
All rights reserved
Printed in the United States of America
Library of Congress Catalog Card Number: 80-80561
International Standard Book Number: 0-88270-430-3
Logos International, Plainfield, New Jersey 07060

To Martha

who kept me on course
and lovingly helped all along the way.

With thanks to Don Tanner
for his timely editorial advice.

Contents

1   A Bite of Eve's Apple   1

2   Mothers for Rent   17

3   It's a Boy!   29

4   The Blueprint of Life   43

5   Planet-Wide Plague   55

6   Building Better Babies   69

7   Fellow Mutants, Arise!   83

8   Quest for Perfection   95

9   The Coming of the Clones   111

10   Closing Pandora's Box   131

11   In Whose Image?   141

12   Redesigned Man   153

Source Notes   167

Let Us Make Man

1

A Bite of Eve's Apple

The rain pattered dismally against the top floor windows, whipped by a cold autumn wind that filled the empty silence of the young doctor's midnight labors. Hunched over the still form before him, he worked slowly and methodically, putting the final touches on two years of effort which had left him sick and pale. Still, in the glimmer of the nearly extinguished candlelight, a fierce obsession could be seen in the tired, fever-dulled eyes—a passion in the depths of his soul that had driven him relentlessly through the grimmest moments of despair.

"Finished at last," the doctor muttered to himself as the elusive spark of life surged into the inanimate giant lying at his feet. In a burst of electrical power, the secret of life was wrenched into the open. Now, thought Victor Frankenstein, a torrent of light would flood into a dark, death-cursed world.

But gazing on his convulsively breathing creation, he suddenly knew it was not to be. The parchment yellow skin barely hid the ugly, rippling muscles and pulsing arteries beneath. Luxuriant black hair and pearly teeth only made the creature more horrible, with staring watery eyes and expressionless black lips.

"I had gazed on him while unfinished," the doctor would write sadly. "He was ugly then, but when those muscles and joints were rendered capable of motion, it became a thing such as even Dante could not have conceived." Bitter and disappointed, Dr. Frankenstein spent the next seven years pursuing the creation to which he had given life, finally exhausting his strength. He died pitifully alone, with his cursed creature still at large.

Mary Shelley's tale of Frankenstein, published in more than seventy-five editions and giving rise to at least thirty films, has struck a responsive chord around the world since its first appearance in 1818. Describing the elusive dream of generating human life—and the destructive misery that can result from tampering in forbidden realms—the story is a gripping saga of the creative drive buried deep within us. Today, for the first time, we are on the threshold of turning that vision into an incredible reality.

Baby From a Bottle

A drab brick hospital in England's grimy industrial heartland was the scene of the dream's abrupt entry into the twentieth century. There, beautiful, blonde-haired Louise Brown, hardly a Frankenstein monster, came crying into the world on July 25, 1978. The first human conceived outside a mother's womb, "Baby Louise" brings us a giant step closer to the day when people can be designed according to a blueprint, developed in a laboratory and born from a mechanical womb.

Whether we like it or not, the genetic revolution has arrived. It came quietly, as gynecologist Dr. Patrick

Steptoe, who had long been attempting to conceive a child in a glass laboratory dish, was approached by John and Lesley Brown. An ordinary working class couple, the Browns were unable to have children because of Lesley's blocked Fallopian tubes (the conduits linking her ovaries to the uterus). They were ready to try anything for a baby.

Having failed hundreds of times in a decade of experiments, Steptoe and his colleague, Dr. Robert Edwards, saw Lesley as the perfect candidate for their procedure. In their small cottage clinic, an egg was surgically removed from the mother, fertilized and grown in the laboratory. When it reached the proper size, the doctors reimplanted it in her womb.

Next followed months of visits during the pregnancy, each of which involved a difficult train trip half the length of England. The couple ran up bills exceeding $20,000, plunging them into debt despite truck driver John's extra jobs and his surprise winnings in a British football pool. Through it all, the Browns desperately hoped to avoid the miscarriage that had plagued Steptoe's earlier patients.

Soon came the happy news that the baby was developing normally. Fearing the controversy that had erupted over similar experiments in the past, the forceful Steptoe clamped a tight lid of secrecy on the proceedings. (In 1974, Dr. Douglas Bevis claimed to have successfully implanted three women who subsequently gave birth. The resulting outcry brought research with human eggs in the United States to a halt.)

Steptoe and Edwards installed theft-proof sliding doors and a red warning-light system in their laboratory, but to no avail. When a story on the pregnancy broke in a local newspaper, the enraged Steptoe screamed to his startled assistants, "One of you is the leak. One of you has talked. I'll find out if it takes me years!"[1]

Tensions mounted as the last weeks of pregnancy approached. Moved secretly to a guarded hospital room under an assumed name, Lesley began the final wait. Her only activities as the dramatic countdown proceeded to the final hour were watching television, knitting and solving crossword puzzles.

Suddenly, the appearance of toxemia in the mother's system forced a quick decision, and Steptoe delivered the baby two weeks prematurely. By Caesarean section, Baby Louise was brought into a waiting world, crying her head off, a beautiful normal baby.

Although the team emphasizes that their work is still experimental, Steptoe enthusiastically predicts that in a few years "it'll be a reasonably commonplace affair."[2] Six months after Baby Louise's birth, a second child, Alastair Montgomery, was born through the duo's efforts (another woman had meanwhile given birth to a test-tube baby in Calcutta, India, independently of Steptoe's work), and hundreds of childless women have since beseiged the clinic. Lesley Brown, too, wants a second child the same way: "I would say to any other woman considering it, 'Go ahead.' "[3]

A Woman's Right?

Hardly had the congratulations died down over this stunning medical achievement than controversy exploded. Several scientists wondered whether all the secrecy was really necessary. Others were troubled by the requirement that the parents agree to abort the unborn child immediately in case of difficulty.

And the questions continue to multiply. Could this possible blessing for millions of families turn into a moral nightmare? "The potential for misadventure is unlimited," says Dr. John Marshall, head of obstetrics and gynecology at Harbor General Hospital, Los Angeles. "What if we got an otherwise perfectly normal individual who was a cyclops? Who is responsible?"[4]

Fired by the challenge of competition, others are pressing for even greater breakthroughs. "If all the pulls and pressures had not been applied," grumbled a UCLA obstetrician, "there might be an American woman now about to deliver [a test-tube baby]."[5] Two years later, his wish is about to become reality.

What was a remote possibility one day is quickly demanded as a right the next. No sooner had Baby Louise come into the world, the first successful birth after scores of heartbreaking failures, than a New York jury awarded Doris Del Zio $50,000 in damages because her hospital stopped an attempt to give her a test-tube baby.

Raising the controversy to new heights of complexity, Mrs. Del Zio asserted that she had lived in mental torment since 1973, when obstetrics director Raymond Vande Wiele of New York's Columbia-Presbyterian Hospital ordered the day-old fertilized egg removed from its incubator and the experiment stopped.

She sued for $1.5 million in damages to "her property," speaking as if the birth of a normal baby were assured from a risky procedure which had been called so slipshod that she might well have died. Although the jury did not award anywhere near that sum, its verdict set an important precedent by finding the hospital at fault, even though her physician, Dr. Landrum Shettles, had deliberately bypassed the hospital's human experiments committee and begun on his own a secret procedure fraught with danger.

What good are restrictions on human life experiments, the hospital director asks, if the court upholds researchers who ignore them?

In hearings across the United States, government investigators have discovered that huge numbers of other infertile women are equally desperate to have babies and will use any technique that might give them the chance. Citing the millions of dollars they pay in taxes, childless couples insist that the government *must* provide them the opportunity for a test-tube baby, risk or no risk.

Ironically, while the strong desire to have a child belongs to the parents, most of the dangers belong to the baby. A typical witness at the hearings, when asked if she would wait a year or two for an experimental procedure to be made safe, thought only of *her* need. "I am thirty-three," she said, "and I don't think I have that much time left. . . . I would be willing to take the chance of [the baby] not being normal. . . . I think I have the right."[6]

Will It Ever End?

From an experiment to a "woman's right" in a few short months—this is just a sample of the dilemma already upon us. "As each new technique leads to another," says biologist Leon Kass of the University of Chicago, "at least one good humanitarian reason can be found to justify each step. The first step serves as a precedent for the second and the second for the third, not just technologically but also in moral arguments."[7]

Notes columnist George F. Will, "Biology is taking mankind into a wild country that is full of threats to the increasingly tentative belief that all human life is of value and should be treated reverently."[8] But according to Steptoe and Edwards, it may be impossible to stop the new procedures because they will be so easy to do. Attempts to legislate against them would only drive the work underground, Edwards insists.[9]

Are we going too far, foreshadowing an era of "womb for rent" and "baby supermarkets"? Have we taken on responsibilities too great for us to handle, requiring wisdom available only to the Creator? Or is this the step that will usher in the last hour of earth?

A New World

One thing is certain: The technical skills have arrived and are growing at an incredible pace. While our knowledge of biology doubles every five years, our understanding of genetics, the master blueprint of all life, doubles every *two* years.

Working with tiny pieces of protein inside a living cell—so packed with information that a teaspoonful would equal a computer one hundred cubic miles in size—researchers make steady progress on new discoveries about life and the creation of new life forms. From a single-celled "bug" that eats oil spills to a black-and-white striped mouse with four parents, the wonders stagger the imagination of a society which considers trips to the moon too commonplace to bother watching on television.

Each advance in biology brings new hope to sufferers of disease. Hundreds of human illnesses are caused by defective genes—the minute specks inside the cells of all living things that control what we look like and how our bodies function. When new techniques of gene transplants and other therapies are perfected, we may see an end to such diseases as cystic fibrosis, hemophilia, sickle-cell anemia, schizophrenia and perhaps cancer. Children will be tested in the womb to determine whether they suffer from genetic disease, which can be corrected by a simple injection, bringing forth a healthy, normal baby.

Other startling applications of the new technology can also be expected in the coming decades.

Totally new types of life will be created, specially designed to solve problems which elude science today.

Intelligence levels could be greatly increased, bringing tomorrow's schoolchildren up to the level of the past's greatest geniuses.

Babies could be born with language skills and multiplication tables built in, much as animals are born with many abilities pre-programmed.

New types of crops can be engineered to help make hunger a distant memory. Grain, fruit and vegetables could be increased in size and designed to contain just the right nutritional value without unnecessary calories. One scientist foresees the development of tablets which would enable people to digest hay and grass, making human food as cheap as fodder. Packaged and marketed properly, the new products might become as popular as hamburger.[10]

Pregnancy Obsolete

And waiting beyond these promises are incredible changes in society advocated today by respected scientists. Conception inside a laboratory test tube could be combined with an artificial womb to allow any woman to have a child, regardless of her health. From emergency use of this mechanical mother, it may be a short step to test-tube pregnancy on demand.

In one view,

> The two tiny laparoscopy scars [from the surgical withdrawal of eggs from the mother], exposed by a bikini on the beach, will be as ordinary as our smallpox vaccination, but women will no longer have lost their figures in childbearing. . . . Motherhood, if it excites any awe at all, will not do so more than fatherhood. . . . [A woman] will find that society does not expect her to have a special relation to her offspring that takes up years of her life, and also she will not expect it of herself.[11]

As the feeling grows that raising children is an expensive burden, this prediction may only await the rapidly developing technology in order to become reality. Women, making up nearly half the work force today, might not want to lose valuable working time by enduring pregnancy and could demand the right to use the new machinery. Once born, the children would be packed off to a day-care center, leaving the parents free to pursue their own interests.

Homosexuals could also enter the reproductive sweepstakes, once they have convinced the courts that they, too, have a right to children. Nurseries of this future world may see carbon-copy babies made of a homosexual father or mother through the process of "cloning."

And since cloning can use genetic material from any cell in the body, other applications could also be found.

Perhaps enough of the chemical composition from the mummy of King Tutankhamen, for example, has survived to allow reconstruction of his living duplicate. Wax museums will be a thing of the past when Xerox copies of the actual people are waiting to greet you at the door!

The same process could preserve nearly extinct species of animals, create a race of Beethovens and Marilyn Monroes, or guarantee physical immortality by storing our duplicates in a cell bank, allowing all of us to be "born again" when our bodies are worn out.

Re-engineering Man

From using genetics for convenience and a better life, it is a small step to a program of re-engineering people to produce intelligence, greater strength and other desirable traits. Dr. Robert Sinsheimer of Cal Tech calls this "potentially one of the most important concepts to arise in the history of mankind. . . . For the first time in all time, a living creature understands its origin and can undertake to design its future."[12]

Already, proposals have been made for determining that future. William Vukowich, associate law professor at Georgetown University, believes that a tax structure could be created to limit deductions for couples with inferior genes and increase the number of deductions for those with superior traits.[13]

Nobel prize-winning chemist Linus Pauling goes further, advocating a tattoo on every young person's forehead, showing his or her genetic makeup to aid in selection of a suitably superior mate.[14] Could this be the forerunner of the world dictator's control system prophesied in the Bible?[15]

Other leading scientists take a dim view of such proposals. "We triumph over nature's unpredictabilities only to subject it to our capricious will and our fickle opinions," observes Dr. Leon Kass. "Thus, engineering the engineer as well as the engine, we race our train we know not where."[16]

The prospect of re-engineering people is no wild fantasy. Much already has been accomplished in this incredible field.

In 1970, two little German girls suffering from a rare hereditary disease were the first human subjects of genetic surgery—an attempt to redesign the cells in their bodies to cure them of the illness.

In 1976, MIT scientists artificially built a complex human gene from off-the-shelf chemicals. Inserted into a cell, the minute gene functioned perfectly, marking a major step toward redesigning man any way we wish.

In 1978, researchers at California's City of Hope used the new technology to turn bacteria into a "factory," manufacturing pure insulin for use by diabetics. Eli Lilly and Company, the pharmaceutical giant, has begun a development program for mass production of the new substance.

Given enough money, predicted Nobel prize-winning geneticist Joshua Lederberg a decade ago, essentially anything we care to do in the area of biological engineering can be a reality within this century.[17] Recent developments have more than confirmed his belief, often coming much sooner than anyone expected.

Perhaps it is only a matter of time before man will have the ability to create a truly living, reproducing being of a type never before seen in the universe. Will we have the wisdom to use this power for the good of humanity? On our answer may hang the future of life on this planet.

The Sorcerer's Apprentice

Intriguing as some of these ideas sound, the darker side of the new technology gives many experts nightmares. The ability to design new life forms is called by some the single most dangerous power to come along in history. Since the required chemicals are readily available by mail order from such companies as Miles Laboratories, makers of Alka-Seltzer, any bright high-school student using a standard biology lab could accidentally unleash an "Andromeda strain" horror. The harmful effects of such an organism, set loose upon an unsuspecting public, might not show up for years, when it is too late to stop its destructive spread.

Dr. Ethan Signer, an MIT biologist, fears that we will create a new organism, "verify a few predictions, and then gradually forget that knowing something isn't the same as knowing everything. We will slowly move from high level containment to low level containment to large scale production to buying the hybrids at the local drug store. At each stage we will get a little cockier, a little surer we know all the possibilities. . . . We are simply asking for trouble."[18]

Some believe that tampering with the human species could mean its end within a few generations, replaced by a new, redesigned race we may not even recognize as human. As knowledge multiplies with increasing speed, many are beginning to question the long-held assumption of scientists that "what can be done must be done."

"It is a hard thing for an experimental biologist to accept," admits Nobel laureate Macfarlane Burnet, "but it is becoming all too evident that there are dangers in knowing what should not be known."[19] Once content to follow their research wherever it led, scientists are now faced with moral choices they may not be equipped to handle.

Looking to the day when diseases can be detected before birth, Sir Francis Crick, a pioneer in the new biology, proposed that "no newborn infant should be declared human until it has passed certain tests regarding its genetic endowment, and . . . if it fails these tests, it forfeits the right to live."[20] Is this where our brave new technology might lead us?

An early example of the growing dilemma was the research of Dr. Daniele Petrucci, who in 1961 claimed to have fertilized a human egg in his laboratory in Italy—just as Steptoe and Edwards would do fifteen years later. Growing it for fifty-nine days, until a heartbeat could be heard, he destroyed the embryo because it became, in his words, "a monstrosity."

Petrucci's work caused an uproar in Italy, and he ceased his experiments when they were condemned by the Vatican. Others, however, were full of praise for the research. The official Chinese Communist press editorialized: "These are achievements of extreme importance, which have opened up bright perspectives for similar research. . . . Nine months of pregnancy is no light or easy burden. . . . If children can be had without being borne, working mothers need not be affected by childbirth. This is happy news for women."[21]

The Russians, too, were impressed, inviting the scientist to spend two months at their Institute of Experimental Biology and sending him home with a Soviet medal. Future experimenters may not be as willing as Petrucci to abandon their work on moral grounds, but at present, little exists beyond individual scruples to halt such research.

Hands Off!

How are people responding to perhaps the greatest social and moral challenge of our time? Scientists generally believe that the research must be allowed to lead where it will, without interference from the public. Proposals to ban further research on human beings, even temporarily, often cause scientists to rally defensively around the flag of Galileo, who was forced to abandon his experiments when they conflicted with medieval church doctrine.

Says Vice President Walter Mondale, "I sense an almost psychopathic objection to the public process, a fear that if the public gets involved, it's going to be anti-science."[22]

Industry, almost totally unregulated in the field of biological research, strongly defends its right to design any new creatures it wishes and take out a patent on them. Dr. Ronald Cape of the Cetus Corporation, the leader in genetic research, insists that practical benefits can come only from free enterprise, with corporate secrets protected by patents, competition as the ground rules, and profits as the motivation.[23]

The corporations maintain that commercial interests and the needs of the marketplace should guide future developments in biology. Sensing a coming breakthrough, they want regulators to stay away. Companies such as Upjohn, Monsanto, du Pont, General Electric, and Standard Oil of Indiana are heavily involved in genetic engineering, which is expected to replace chemicals in importance by the end of the century and grow into a multibillion-dollar industry.

The U.S. government has generally taken a hands-off approach to genetic research, enthusiastically funding many of the new discoveries but allowing the scientists to follow their own interests with only a few restrictions.

Behind government thinking always lies the threat of international competition and the desire to keep our options open. Today, the leader in much of genetic research is Great Britain; tomorrow, we may be in a desperate gene race with the Soviet Union. Fears of a "gene gap" have already been raised in U.S. military publications, while the Soviet Communist Party issued a decree in 1974 calling for expansion of genetic research, regardless of cost.[24]

"Sooner or later," says one authority, "one human society or another will launch out on this adventure, whether the rest of mankind approves or not. If this happens, and a superior race emerges with greater intelligence and longer lives, how will these people look upon those who are lagging behind? One thing is certain: they, not we, will be the heirs to the future, and they will assume control."[25]

Crossroads of History

We stand at the crossroads in human history, and few seem willing—or able—to decide which way to go. Faced with the greatest moral and religious challenge of our lifetime, all of us have a vital need to be informed about the present progress and future possibilities of this awesome science, ready to take action before our destiny is sealed for us. The decisions we make today will have profound implications for the future of the family, marriage and the meaning of life.

Perhaps this new technology has been placed in our hands as the final step in restoring the perfect mind and body once possessed by Adam. But will we pervert it into a new Tower of Babel, lifting ourselves to unimagined heights of arrogance and rebellion?

Can we consider abandoning research which holds out hope to so many millions of suffering humanity, just because it might be misused? If science can discover defects in an unborn child but cannot yet correct them, what are the parents to do? Consent to an abortion? Bring into the world a baby they know will be severely retarded or doomed to an early, painful death? Or refuse to allow any tests at all and blindly hope for the best?

We may be able to create artificial life in a laboratory, but what would be the purpose? Would this set the final scene on the world stage, moving the Author to call down the curtain?

The questions are many and the answers difficult to find. The incredible possibilities that arrived with Baby Louise present an awesome challenge to our most cherished beliefs. Where will we go from here?

2

Mothers for Rent

Early in 1977, readers of several Michigan college newspapers were shocked to see an advertisement placed by a married couple identifying themselves as "Al and Betty." Betty was prone to miscarriage, the ad explained, and the couple sought a woman who would agree to be artificially impregnated and carry their baby to term.

A number of young women responded, asking fees close to $5,000 for this service. Under the arrangement, Al and Betty received a child, and the donor earned enough to cover college expenses for perhaps a year.[1]

This bizarre case is far from unique. Similar advertisements have appeared in newspapers across America, with womb-for-rent prices said to hit $20,000 in major cities. In California, the traffic in babies has become so lucrative that police suspect a growing Mafia involvement.

Even actor Richard Burton joined the market a few years ago. Advertising in a British magazine, he offered $25,000 to any woman under thirty-eight who would bear him a girl and a chauvinistic $50,000 if it turned out to be a boy. The response he drew went unreported.[2]

In an era of well-publicized shortages, the lack of adoptable babies is one rarely discussed, yet it means a lot to the one couple in eight with fertility problems. Because of the Pill, widespread abortion and the desire of most unwed mothers to keep their babies, a childless couple might have to wait seven years and undergo extensive investigation before a suitable baby is offered to them—if one is available at all.

Unwilling to endure these lengthy adoption procedures, couples are turning to the "gray market" in babies, and increased demand has made the purveyors of infants less scrupulous than ever. Shocking stories have come to light about baby farms at which young women bear children, intending to sell them for adoption. The going rate today ranges between $10,000 and $25,000 per infant. One operator is reported to have netted more than a million dollars a year by flying pregnant Eastern European women to the Caribbean, where they enjoy a free holiday and are paid $3,000 each for their babies.[3]

With the arrival of the biological revolution, however, these extraordinary measures will seem crude and old-fashioned. New techniques on the horizon will allow any couple who wants children to have them, and select their sex and physical features in the process! Natural or artificial conception, development within the mother's womb or outside, breeding babies with the genes of one parent, both or neither—the coming possibilities stagger the imagination.

Hired Mothers

A refined version of the womb-for-rent, called "surrogate motherhood," will use some of the skills that Steptoe pioneered with test-tube babies. In this procedure, the child is truly the product of both its father and mother, but is carried to term by another woman as a human incubator. In the near future, a doctor will surgically remove an already-conceived baby from the natural mother's womb and transplant it into another woman—a sister, friend or hired mother. There, the child will develop to term and be born into the waiting arms of its real mother.

Who will be the clients of these substitute mothers? Thousands of women cannot carry a baby because of poor health, and many others are prone to miscarriage. With a hired womb, they too will be able to experience the joys of parenthood as the new biology provides hope for thousands who otherwise would be forced into the frustrating and sometimes heartbreaking adoption process.

Surrogate motherhood might be an attractive, full-time profession for a young woman who is divorced or widowed, especially if she already has children at home. Good health and a relaxed life style might be the only requirements for such a career. Some see it as a way to reduce welfare payments to mothers with young children, allowing them to earn enough during their childbearing years for job training later.

Undoubtedly, foreign competition would soon enter the picture, driving the price down unless a "Baby-bearers' Union" passed tariff restrictions or forced the licensing of these human incubators. Still, suitable women could always be found at a reasonable price, perhaps among the population of illegal immigrants.

Some will seek a hired womb merely for convenience. A career woman who fears the loss of income during pregnancy, for example, or a movie star who dares not lose her figure might be clients. As women's liberation and self-interest travel a parallel course toward this new science, the availability of substitute mothers might well be demanded as a woman's right.

What kind of mother would a woman be who is unwilling to inconvenience herself during the few months of natural pregnancy? Some people believe that such a person should never be allowed to have children, fearing that the infants will be psychologically scarred and starved for affection.

Others counter that rental motherhood is no worse than the wet nurse, a common sight until fifty years ago when baby bottles and infant formulas came on the market. Might not a future generation casually accept surrogate mothers as an updated version of the wet nurse?

A Worldwide Business

The transplanting of a tiny embryo from one womb to another may seem like a distant dream, but already it has become big business with livestock. Several companies are reaping hefty profits by flushing just-conceived embryos from superior breeds of cattle and implanting them in ordinary cows, which carry the offspring to birth. These super-calves can command twenty thousand dollars or more, with less exotic breeds averaging three thousand dollars.

Recent developments are opening up other new vistas, including the ability to send cow embryos around the world as part of a mass marketing system. One method is to temporarily store them in the Fallopian tubes of a living rabbit for air shipment to the customer, who can then reimplant them in his own cattle. Success rates as high as ninety percent are predicted with this method.

More attractive still is the transplanting of *frozen* embryos, held in storage until required by a breeder. Major breakthroughs in the 1970s have enabled researchers to freeze the embryos of mice, rabbits and cows in liquid nitrogen for as long as five.years, with a high percentage of successful births after thawing and implantation. With the perfection of this technique, great improvement in livestock is expected worldwide.

Will embryo transplants—frozen or not—be applied to humans? Dr. Kurt Hirschhorn of Mt. Sinai School of Medicine says, "There is no question whatever in my mind that all of this is going to happen."[4]

The science of embryo transplanting can work both ways. Not only could an unborn child be moved into a donor womb, but a donated embryo could be made available to a couple unable to conceive a baby of their own. If both husband and wife are sterile, or are carriers of a hereditary disease, they could employ a carefully selected donor couple of the same ethnic, social and religious background whose embryo would be implanted into the wife. She would carry the baby as her own, and happily give birth to a child no one suspects is "adopted."

Even more futuristic is the idea of an embryo supermarket. A couple of the next century could walk into the corner drugstore and select their offspring from a shelf carrying a dazzling array of frozen (or freeze-dried?) embryos in little packets. Each would be carefully labeled, identifying sex, race, eye color and probable IQ of the resulting child. Guaranteed to be free of genetic disease, the embryo could be purchased and taken to the doctor for implantation—if the shoppers are able to choose from the overwhelming selection.

No longer will parents be forced to take whatever kind of child comes along. Having the ultimate in consumer power, they will be able to select the composition of their own families.

A more radical idea is to implant human embryos, not in substitute mothers, but in *animals*. Taking availability into account, embryologist Robert Francoeur sees small cows as the best choice. Gentle, sedentary, and having a gestation period which matches the human nine months, cows could carry a human baby until near term, when it would be transferred to an incubator for the last few weeks before birth.[5]

Naturally, a youngster might resent having spent eight months in a pasture while his mother was off giving parties. That is just one of the psychological, legal and moral questions raised by the whole subject of embryo transplants and mothers-for-rent.

For instance, if a hired mother decided she wanted to keep the baby, who would be considered the real mother? Could the substitute mother be held responsible for defects in the child? What about the youngster's risk of a severe identity crisis, more traumatic than the present-day agonies of adopted children wondering who their real parents are?

Mothers for Rent

Baby Under Glass

Before answers to these questions can be found, the new biology will have moved on to a substitute for the substitute mother.

For several decades, doctors have been able to save premature babies by placing them in incubators, and the threshold of survival is being steadily pushed back. In 1979, for example, little Mignon Faulkner was born at only twenty-three weeks. Twelve inches long and weighing a mere seventeen ounces, she was saved because of modern incubation methods.

Now, scientists are also working forward with incubators from the time of conception, looking to the day when the two directions of research meet and a baby can be carried from conception to birth in a womb of glass and steel. Artificial wombs are now under development in Britain, America and the Soviet Union.

Among the techniques being tested is one by Dr. Robert Goodlin of Stanford University, who has put unborn babies in a thick-walled steel chamber containing a viewing porthole. Bubbling oxygen at immense pressure into a saltwater solution, the mechanical womb drives the life-giving gas directly through the baby's skin, sparing the inadequately-formed lungs the need to work.

Other promising experiments have been performed with animal and human embryos, but the major remaining problem is disposing of the carbon dioxide and other chemical wastes that build up within the artificial womb.

Reportedly, the Russians are far more advanced than their Anglo-American counterparts in this field. Led by the brilliant biologist Dr. Petr Anokhin of the

Institute of Experimental Biology at the Academy of Medical Sciences, Soviet researchers claim to have kept several hundred human babies alive in artificial wombs for up to six months.[6]

According to the respected British journal *New Scientist*, it is only a matter of time before a baby is carried to full term inside a mechanical womb.[7]

Of course, great difficulties lie ahead. The placenta—a membrane that surrounds and protects the baby in the womb—is one of the most amazing and complex of biological marvels. Screening out harmful substances, it acts as a filter to pass chemicals, oxygen and sugars vital to life from mother to child. Scientists labored five years just to understand how oxygen and carbon dioxide are carried across the placental membrane; full development of the mechanical version will involve an enormous amount of research.

A New Liberation

What would be the point of such an effort? In an article titled "The Obsolescent Mother," Edward Grossman lists several reasons. A mechanical womb would:

1. Enable technicians to detect and treat disease before birth;

2. Provide a safe, controlled environment for the developing baby;

3. Allow women who cannot carry a baby to have children;

4. Pemit doctors to take tissue samples and freeze them for organ transplants later in life.[8]

Each one of these is considered by many scientists to be sufficient reason for developing the artificial womb.

Like surrogate motherhood, baby-under-glass technology can also be used merely for convenience, although
it would be expensive convenience at first. In Russia,
the freeing of women from pregnancy might be worth
the expense, though, because eighty percent of the
working age women have jobs—and need them to
make ends meet. Perhaps the same will be true in
Western countries before long, turning convenience
into necessity.

Still, the idea of growing a tiny human inside a
machine makes many uneasy. With all the proposals
for manipulations on unborn children—increasing
brain and head size, changing sex, eye color or other
basic attributes at will—the introduction of an
artificial womb might turn these modification schemes
into reality. At the very least, "motherhood" would
undergo a radical change, and probably not for the
better.

From the women's liberation point of view, however,
the new science provides the ultimate in freedom. "It
seems apparent," speculates one author, "that many
women . . . would avail themselves of the [artificial
womb] as soon as it were offered, eagerly foregoing the
rigors of pregnancy by dropping off at the baby
factory their two-to-three-day-old embryos. . . .

"Along the nine-month way, they could drop in at the
laboratory to check on their baby's progress, perhaps
ordering a few changes here and there. The 'coming
out' party could well take on a whole new meaning as
proud parents and their friends gather at the mouth of
glass-and-steel wombs for the 'birth' of thier babies."[9]

Vanguard or Victim?
From the baby's side, the idea may not look so rosy.

Despite all the talk about how unsafe a mother's womb is, compared to the as-yet-unproved mechanical imitation it is a fortress of security. Warns author Gerald Leach, "If there is one thing biomedicine has learned about the embryo in the last twenty years, it is its extreme and usually disastrous sensitivity to a long list of chemical agents, or their lack."[10] Some believe that the physical and mental defects which threaten "decanted" babies explain why the Russians have not announced the artificial womb breakthroughs that were expected.

And what about the psychological damage a mechanical gestation could do to the child? In the natural womb, a baby experiences cycles of sleep and activity, senses emotion and human voices, kicks against the resilient placenta and feels the reassuring surge of a maternal heartbeat. How could these be simulated in a machine?

Above all, whether dealing with artificial wombs or human substitute mothers, the most important question concerns the spiritual bond between mother and child, indefinable but crucial. As science becomes increasingly mechanical in outlook, the natural womb is seen as merely a complex machine to keep the baby nourished and safe while it develops. But how can we be sure?

New levels of relationship between people are discovered constantly. If a person can suffer anguish when his identical twin dies half a world away, as has been reported many times, what might be the hidden bonds of intimacy between a mother and her unborn child? Something as simple as breast-feeding seems to enhance the baby's psychological adjustment, according to many experts. Certainly, the fetal relationship is much more fundamental.

Mothers for Rent

More than just a modern wet nurse, the substitute mother—mechanical or mercenary—is a harsh intruder in a subtle relationship that goes beyond simple nurture. If we use a genetic mother to produce the egg, a natal mother or machine to carry the child, and possibly a social mother to raise him, aren't we ripping apart a God-created union and robbing the child of a spiritual bond he desperately needs? Our economy may provide this new service, but it is doubtful whether society, the marriage of the couple involved, or the psychological and spiritual balance of the child would be improved.

Yet despite doubts, some are ready to forge ahead. The widespread prediction among scientists is that the first artificial womb will be a reality within twenty years, while substitute motherhood awaits only a doctor willing to perform the embryo transplant.

Says one American scientist who insists on anonymity, "If I can carry a baby all the way through to birth *in vitro* [in glass], I certainly plan to do it—although, obviously, I'm not going to succeed on the first attempt, or even the twentieth."[11]

When the artificial womb is ready to take its place alongside Steptoe and Edwards's work in laboratory conception, the complete test-tube baby envisioned in Aldous Huxley's *Brave New World*—growing from conception to birth inside a bottle—will be a reality. When Huxley wrote his famous book in 1931, he considered this fearful day to be perhaps six hundred years in the future. It is a measure of the new biology's blinding speed that we are already on the threshold of that moment, and many other incredible powers are waiting in the wings.

3

It's a Boy!

An Atlanta woman recently visited the office of Dr. Desider Rothe at New York's Cornell Medical Center with a bizarre demand. "I'm pregnant," she announced, "and unless you can prove to me that the baby will be a boy, I'm having an abortion."

Wanting to save the child, the doctor took a sample of the fluid that surrounds the baby to analyze the cells it contained. This procedure reveals the child's sex as well as the presence of disease. To Dr. Rothe's sorrow, the unborn baby was a girl, and the woman had it aborted.

Several months later, she reappeared in his office, and the ugly scene was repeated. The discovery of another girl led to a second abortion.

On her next visit, the woman learned that this third child was a boy, and she allowed it to live. For her, it had been worth two abortions and the cost of three round-trip plane tickets from Atlanta to New York just to guarantee a male offspring.[1]

The brutal destruction of babies of the wrong sex— long a part of human history—will soon give way to startling new powers of selecting the sex before the child is conceived. Several popular books are on the market today which outline a controversial means of "natural" sex selection, based on physical indications like those used in the rhythm method of birth control. Scientists are working on more sophisticated procedures which may bring sex selection to the general public within a few years.

While all human eggs are similar, some researchers believe that the sex-determining sperm are slightly different from each other, depending on whether they would produce a male or a female. Excited by this discovery, they believe that reliable methods for separating the two sperm types—and thus deciding the sex of the children—are just around the corner. A few institutions, including Michael Reese Hospital in Chicago, have already put selection techniques into practice on an experimental basis.[2]

Of at least five different approaches under study, two show the most promise. One is to immobilize the sperm cells by refrigeration, allowing the heavier, female-producing ones to settle to the bottom of a test tube. The other technique involves passing the sperm through an electrically charged resin which will catch the male-determining cells and let the others pass. The two methods can be combined for a higher success rate, with very close to one hundred percent reliability predicted for the future.

Because it is a spinoff from lavishly funded research on fertility and contraception, and since no major obstacles are seen, scientists believe that sex selection will soon be available to all. Australia's Dr. Charles Birch foresees production of a sex-selecting pill that prospective parents can buy at the drugstore—pink for a girl and blue for a boy.[3]

Boys, Boys, Boys

Will people use this incredible new power? What would be its impact in a world that maintains a delicate balance between the numbers of men and women?

Look at the advantages. No longer would a girl be shunned because her parents wanted a boy, and all of us could plan a family exactly as we desired. Genetic diseases would lessen, because many of them are sex-linked. (For example, a mother who is a known carrier of hemophilia [bleeder's disease] would choose to have girls; the illness appears only in male offspring.)

Dr. Paul Ehrlich, the famous population expert at Stanford University, has called for increased research on sex selection. He believes that many couples would limit their families to two children if they could be assured of having a boy and a girl.[4] In underdeveloped countries, especially, the pressure to produce sons would no longer force ever larger families, a fact which might make their governments very interested in this technology as part of population control.

If people could select their children's sex, which would they choose? Study after study yields the same result: People want boys more often than girls, especially as their first child.

In a 1970 Princeton University survey of six thousand women of childbearing age, thirty-nine percent said they would choose their child's sex if they had the chance. A 1974 study of married couples brought that percentage up to forty-seven, and it will certainly go higher when the procedure is actually available.[5]

Of those who would make the selection, ninety percent wanted a boy first, and if limited to only one child, seventy-two percent would choose a male. Some experts believe this would raise the male-female proportion from its present 51.5 percent male births to a wildly unbalanced sixty or seventy percent.

At the very least, this would give most girls an older brother, and since firstborn children usually are the ones to excel in later life, sex selection would further increase male dominance. Male and female differences would be emphasized as parents select one over the other, building a new legacy of conflict between the sexes.

Columbia University sociologist Amitai Etzioni points out that men are much more likely to become criminals than women, while fewer of them are churchgoers or involved in the moral instruction of children.[6] Beyond the obvious problem of finding mates for the larger number of men available, some futurists see the return of a violent, frontier-type society of unrestrained male dominance.

The irony is that women themselves say they would choose to have boys at the very time when progress is supposedly being made in building the female self-image. Dr. Elizabeth Connell of Planned Parenthood speculates that "American women feel more female giving birth to males, thus pleasing their husbands," and the National Right to Life Committee's Dr. Mildred Jefferson warns that this new technology might "liberate women out of this world."[7]

Because sex selection research is virtually unstoppable, a few scientists want laws to limit it to those with sex-linked diseases like hemophilia. They would prohibit doctors from telling parents the sex of their child as determined by prenatal tests except for this reason.

This approach or one similar might be worth careful consideration. Why should we play God in ignorance of the consequences, contributing to human sorrow?

In the majority of cases, sex selection would be a matter of personal taste, not necessity. Is such basic tampering with the next generation warranted by the whim of parents?

Bioethicist Marc Lappé concludes: "A society which would choose male offspring in preference to females (or vice versa) is probably not one which is ready for the responsibility of assuming regulation of the balance between men and women. Certainly, one which would preferentially deny rights and opportunities to women would appear to disqualify itself as having the necessary fairness and wisdom to proffer the option of sex selection."[8]

Only Her Doctor Knows

Whether the goal is sex selection, womb-for-rent or overcoming fertility problems, many of the new wonders use artificial insemination, the fastest-growing of any biotechnique, to achieve pregnancy.

Basically, this means collecting and concentrating sperm cells, which are then inserted into the womb by a doctor using a high-powered syringe. A similar process using a woman's eggs, called artificial inovulation, or egg grafting, has also been developed. Because of its greater complexity, it is much less common so far.

Artificial insemination has been used over the years by many thousands of married couples to compensate for the husband's low fertility levels or other difficulties in reproduction. Success rates, however, are disappointing—twenty percent or less—because male fertility problems usually have a combination of causes, which mere concentration of sperm may not overcome.

Much more widespread—and controversial—is artificial insemination by an anonymous donor (called *AID*). Usually, these are university graduate students who have been carefully screened to rule out the possibility of their carrying genetic disease. Although only three percent of the American public had ever heard of AID a decade ago, it is estimated that by now half a million of their fellow citizens may have been conceived in this way, with twenty thousand more coming each year.

Assuming a pregnancy rate of twenty percent, that would mean up to one hundred thousand AID attempts each year in the United States, with its popularity in Europe at least as high. Clearly, AID is a major social phenomenon, and with adoption becoming more difficult, we can expect a heavy increase in the years ahead.

Artificial insemination experiments with humans and animals began hundreds of years ago, and the process has now become dominant in animal breeding. Today, ninety-five percent of all cattle born in the United States are conceived artificially, totaling more than 100 million to date, as are additional millions of sheep, pigs, goats and turkeys. The business continues to expand as shipments of frozen sperm are sold to breeders all over the world. (A recent case concerning the smuggling of prize bull sperm into the United States went all the way to the Supreme Court. At stake was nearly a million dollars in retail value.[9])

With humans, AID has become almost as routine, costing between $75 and $100 a visit. Donors remain strictly anonymous in most cases, and quite often the obstetrician is not aware when it has been used.

At least half of the infertile couples in a 1967 survey indicated they would prefer AID over adoption, for several reasons. A far simpler and less expensive procedure, it allows the child to have the genes of one parent instead of neither and eliminates the fearful risk of the natural mother reclaiming her child before adoption is finalized.[10] Since the physical characteristics of the donor are carefully matched to the husband's, the children look "natural," and even snoopy relatives would never guess the truth.

As the public becomes more aware of AID, it can be expected to increasingly accept the practice. Says Dr. Wayne Decker of the New York Fertility Research Foundation, "A lot of things we wouldn't do a few years ago, we no longer think twice about. For instance, I do forty or fifty artificial inseminations a week, whereas a few years ago we would do ten or twelve a year. The repellent connotations of artificial insemination are almost nonexistent now. Couples not only accept it, but seem to regard it as more natural than adoption."[11]

Greater numbers of lesbians are approaching doctors with the intent of conceiving children through AID, and many physicians are quite willing to help. Janis Hetherington, a thirty-three-year-old antique dealer, was the first homosexual in England to have an acknowledged AID child. A thin woman with short-cropped black hair, Janis appeared on the NBC television program, *Weekend.*

"Having sex with a man in order to produce a child would be a lie," she insisted. "Still, I've always wanted a child." Living with her female lover and the woman's adopted daughter, Janis sees herself and son Nicholas, now eight, as part of a happy family. She hasn't had the slightest reluctance to tell the boy the details of his birth, either.[12]

Similar stories can be found in the United States. Asserts a young Southern California lesbian, "Just because I'm gay doesn't mean I don't want to experience pregnancy and parenting. I've made my sexual decision; now I should have the right to decide if I should bear children. Heterosexual women have the right. Why shouldn't I have it too?"[13]

Dr. Decker, who freely admits making AID available to lesbians, defends the practice. "I'm not a moralist. To me, the kind of family community the child will be born into is more important than the mother's sexual preference."[14] Some would counter that sexual preference has a great deal to do with the kind of family the child is raised in.

Deep Freeze Banking

Today, artificial insemination, both for married couples and single parents, is beginning to receive a substantial boost by another new science: frozen sperm stored in sperm banks.

The first baby conceived from frozen sperm was born in 1953, a healthy boy profiled at age seventeen in a *New York Times* article as a normal teenager who earned mostly A's in school.[15] Since then, thousands of others have been conceived in this way, with the numbers growing all the time.

Commercial sperm banks are open in a dozen American cities to provide both the service of storing a husband's sperm and of making the sperm of donors available for AID. At one of the largest, Cryogenic Laboratories, sperm can be deposited for an initial fee of fifty-five dollars and an annual storage payment of eighteen dollars.

Gradually frozen by liquid nitrogen to a temperature below -300° F., the sperm is stored in a thermos-like container about the size of a cigar tube. Because one sample in four fails to survive freezing, a small part of every deposit is thawed and checked to determine its potency. No one knows how long frozen sperm can survive, but babies reportedly have been born from sperm frozen as long as thirteen years.[16]

In the early 1970s, it was assumed that husbands undergoing vasectomies for birth control—more than a million a year in the United States—would provide a natural market for frozen sperm banks, but that has not been the case. Very few of these men intended to change their minds and consider children in the future, so most of the early banks went out of business.

The remaining firms have turned heavily to artificial insemination as a primary source of income, providing frozen sperm from a broad selection of coded donors. One company, Idant, has more than ten thousand samples in storage, complete with details about the background of each donor.[17]

Anonymity is much better protected in large commercial sperm banks than in private doctors' offices, and as the practice becomes more widespread, such companies are likely to be the normal source for AID babies. Perhaps an international market will develop, and parents—like cattle breeders today—will be able to order frozen sperm from a catalog.

Unfortunately, the seed bank customer must rely completely on the integrity of the firm being dealt with, and such trust is not always warranted. One San Francisco depositor discovered after his vasectomy that the sperm bank had accidentally destroyed his specimen. He sued for five million dollars.

George Washington University's Mark Frankel has studied frozen sperm banks and found an alarming lack of standardization in the business. Donor screening is haphazard, and knowledge about the best methods of storing specimens is limited. People are operating on blind faith, he concludes, when they trust the heredity of their children to commercial sperm banks.[18]

And what about the companies that went out of business so suddenly? In a typical case I know of, the initial fee was paid and sperm deposited just before a vasectomy; the company was never heard from again. Isn't this a slipshod way to deal with a family's future?

Still, the banking continues to increase, leading some visionaries to foresee the day when all young people will make deposits in a sperm or egg bank after puberty and then be sterilized to guarantee society-wide birth control. When a child is desired, the husband and wife would draw on their respective deposits and conceive a baby, perhaps by license from the state.

With frozen sperm banking, even the death of one of the partners does not rule out building a family. In a celebrated 1977 case, cartoonist Kim Casali, who draws the popular "Love Is . . ." cartoon strip, gave birth to a son sixteen months after the father's death from cancer, the result of artificial insemination by his previously frozen sperm. The biologist's imagination is the only limit to what can be done with the amazing new discoveries.

AID or Handicap?

Widespread as AID has become, many people are deeply concerned about the psychological and legal problems involved. Some studies indicate that AID can be harmful to marriage—especially when wives are obsessed with curiosity about the donor.

It is noteworthy that very few couples ever repeat an AID birth. One husband remarked, "I couldn't go through that again. If the child can't be mine, I don't want it to be either of ours."[19]

And the children conceived in this manner? At present, virtually none of them ever know they have been born of AID, and psychologists want to keep it that way. "The identity problems of such a child could be enormous, and perhaps devastating," warns psychotherapist Annette Baran.[20] Thus, the parents are forced to live with a secret they can never openly discuss, a conspiracy of silence that may strain family relationships to the breaking point.

With AID conception hidden, a slight risk also exists of half brothers and sisters—conceived from the same donor—unknowingly marrying. In a 1974 incident reported by *Time* magazine, a doctor had to step in at the last moment to prevent the incestuous marriage of two young people he knew to be related through a common AID father. Some clinics now limit each donor to one hundred conceptions to reduce the odds of such an occurrence.

Legally, AID can be a nightmare. For example, a child born of artificial insemination is legitimate in California, illegitimate in Illinois and legitimate in Oklahoma only when adopted by the mother's husband. Doctors in some states are uneasy about filling out the birth certificate with the husband's name, knowing he is not the father. In case of divorce, the legal wrangling can become extremely complicated. One New York husband even sued for divorce on the grounds that the doctor who performed AID on his wife had technically committed adultery.[21]

Most important, what are the moral issues involved in artificial insemination? Is AID adultery, or merely one step removed from adoption? Since the procedure is performed in secret, each couple considering it should have the question settled in their minds, because they will most likely be making this crucial moral decision for themselves.

Nearly all ethicists and religious bodies—with the exception of the Roman Catholic church—accept artificial insemination by the *husband* as morally permissible. Aside from the physical difference in conception, there seems to be no ethical problem involved in helping nature along within the bonds of marriage. (One caution should be noted, however: Most problems of male impotence have a psychological, not physical, origin. The arrival of children may make existing marital difficulties worse.)

On artificial insemination by donor, attitudes have been changing over the past few decades. While almost universally condemned by church leaders in the 1940s and early 1950s as being a form of adultery, it now has the approval of many Protestant religious leaders.

With a completely anonymous donor and the full consent of the husband, they reason, how could it possibly be considered infidelity to the marriage convenant? And since no emotional tie to the donor exists, and the procedure takes place as an act of love within marriage, they see no sexual sin involved.

Some of the more liberal ethicists, such as Joseph Fletcher of the University of Virginia, don't bother about the question of sexual fidelity at all. "It is a mistake ethically and humanly to equate being faithful in marriage with making it a sexual monopoly," he insists. "No longer can we say that a monogamous marriage agreement means exclusive access to any or all of the wife's or husband's 'generative faculties.' "[22] Instead, marriage is seen as a convenient institution to contain family love and commitment; the exact source of sperm and egg is secondary.

What does the Bible say about these views? The advocates of AID often claim a biblical basis for the practice, pointing out that Sarah gave Abraham her maid in order to supply the male heir promised by God.[23] AID, they insist, is a less adulterous example of the same principle.

They fail to note, however, that Abraham was following the common practice of the pagan culture in which he lived, rather than obeying any command of God.[24] In fact, the child of that union was cast out by divine decree, because Abraham had been wrong in attempting to have a son outside his marriage to Sarah.

The Bible considers marriage a covenant of two becoming "one flesh."[25] Attempts to split off the physical union and call it a mere mechanical function, as the advocates of AID suggest, violate the biblical meaning of parenthood in order to make parents.

The fact that AID is impersonal does not change the implications: prostitution is also anonymous and impersonal. In giving grounds for divorce, the Bible does not question the motives of the partner who is sexually unfaithful; the physical act is what rips apart the fabric of the marriage.[26] Even if the donor were an identical twin, with exactly the same genes as the husband, the marriage union would still be violated, because the covenant exists between two human individuals, not two sets of chromosomes.

As science makes new wonders available, each of them will have a moral impact that affects our lives at the deepest levels. Man has long dreamed of the day when he could control birth and create life, and now we are on the edge of that era. But it did not arrive overnight, and the tale of its development foreshadows the new biology's incredible significance.

4

The Blueprint of Life

Among the mysterious and magical alchemists of the Middle Ages, none was more controversial than the brilliant Swiss who called himself Paracelsus. Like his fellow wizards, the famous doctor spent hours in his laboratory, trying to unlock the greatest of all mysteries: fashioning life according to a complicated magical formula.

The creation of an artificial being, the medieval master wrote, is "one of the greatest secrets which God has to reveal to mortal and fallible man, which deserves to be kept secret until the last times, when there shall be nothing hidden, but all things shall be made manifest."[1] Although he failed to produce the miniature man his recipes promised, Paracelsus kept the dream alive, inspiring generations of storytellers to come.

About the same time as Paracelsus, the High Rabbi of Prague, Loew ben Bezaleel, was said to have given life to a shapeless mass of clay called a "golem." According to this popular Hebrew legend, the rabbi wrote the secret name of God on a piece of parchment and placed it on the hulk's forehead.

Suddenly endowed with the gift of life, the monster

ran wild, turning on its creator and causing great destruction. When it finally profaned the Sabbath, the rabbi sadly realized it could do only evil.

Variations of the golem legend describe creatures formed from red potter's clay, no taller than a six-year-old child. Energized by God's name, they were at first small, docile slaves who worked willingly and never tired, but they soon grew into towering masses of solid stone, sometimes crushing their creators underfoot. Although not basically evil in themselves, the golems wrought terrible havoc because their power could not be controlled.

The theme turns even more sinister when we enter the twentieth century, where the idea of tinkering with life is represented by Huxley's *Brave New World*. In this chilling vision of the future, children are grown in factory-like baby hatcheries, with fertilized eggs multiplied into myriads of clones.

As the director of the Central London Hatchery explains, "One egg, one embryo, one adult—normality. But a bokanovskified egg will bud, will proliferate, will divide. From eight to ninety-six buds, and every bud will grow into a perfectly formed embryo, and every embryo into a full-sized adult. Making ninety-six human beings grow where only one grew before. Progress."[2]

Relying on the hatcheries, the masters of Huxley's society are able to create armies of standardized men and women, fabricated in uniform batches. "Ninety-six identical twins working ninety-six identical machines!" the director exclaims enthusiastically. The ultimate in social control.

Through the ages, whether in the older setting of a

lone dabbler working in his laboratory or the more modern scene of an impersonal, rigidly controlled factory, the heart of the vision is the same. Man tries to step beyond his mortality, taking over the very forces to which he owes his existence. But the creation can be no more perfect than the creator, and the stories always end with the creatures going out of control, like the runaway forms animated by a bumbling sorcerer's apprentice. Perhaps these dreamers of the past have a warning that the age of biotechnology might well consider.

Have we now reached the last times anticipated by Paracelsus, when no secret is hidden from us? How did we arrive at this awesome height of knowledge, turning the fantasies of medieval wizards and creative story-tellers into the hard realities of laboratory science? To find answers, we must first step back to a quiet monastery garden in nineteenth-century Austria, where a young monk made a remarkable discovery.

The Monk's Garden

The child of a poor peasant family, Gregor Mendel struggled to support himself through long, difficult years of education, nurturing all the while a lifelong interest in plant life. Pale, half-starved and subject to severe depressions, the brilliant young man entered the Augustinian order after graduation from the University of Vienna. He looked forward eagerly to the quiet life of a monastery, where he could continue his botanical studies in the garden.

A gifted teacher, Mendel was soon popular with both students and colleagues in the little high school near Vienna to which he was assigned. His tall hat, frock coat and high boots with tucked-in trousers quickly became a fixture in the district. Pupils constantly flocked to this round-faced, stocky little man who peered at them good-naturedly through gold-rimmed spectacles.

But teaching was not to be his contribution to history. Rather, his small garden behind the monastery—one hundred feet long and about twenty feet wide—became the site of crossbreeding experiments with flowers and peas which would form the cornerstone of modern genetics.

"Why," he wondered during his experiments with ordinary garden plants, "would a cross between round and wrinkled peas yield only round ones? And yet, if those round peas were in turn planted, they would produce both round and wrinkled peas once again—often in the same pod." Puzzled, he concluded that the hybrid pea seeds must be different inside, and he began a search for that inner secret.

Tracing a few features of the pea seeds, such as round or wrinkled, white or gray, Mendel found to his delight that the results were not a muddy mixture of traits but a clear dominance of one over the other. When tall plants were crossed with short ones, for example, the offspring were always tall. The prevailing trait is called *dominant* and the masked trait *recessive*.

Excited by his findings, the monk worked patiently to produce a third generation of the hybrid peas and made an even more remarkable discovery. The recessive traits, which had been completely masked in the second generation, not only reappeared in some of the offspring, but in almost the exact ratio of three plants with dominant characteristics to one with recessive.

No matter which trait he followed—tallness, color, wrinkles—seventy-five percent showed the dominant character and twenty-five percent the recessive.

Through this simple experiment, Mendel discovered that cells for reproduction are different from all other living cells. They divide in two, with each parent supplying half the traits that appear in the offspring.

Inside the cells are substances, which we now call genes, responsible for passing along all the hereditary characteristics that make families alike and give each of us our unique features. These genes are arranged in a line on tiny, thread-like strands called *chromosomes*, which operate in pairs.

When a sex cell is formed, the chromosome pairs separate, one of each duo going to the new cell and the other remaining in the original. Every chromosome carries a portion of the blueprint that makes you "you." When the appropriate chromosome from the father is paired with the one from the mother, only the dominant traits appear in the children. Recessive traits show up when, by chance, two recessive chromosomes happen to pair off. This occurs on the average of one out of four times, accounting exactly for Mendel's observations.[3]

The Elusive Gene

Understanding at last that genes transmit heredity, scientists now had another great hurdle to cross before they could begin tinkering with the building blocks of life. They had to find out what genes were.

The discovery was made a century after Mendel by an unlikely duo of bold young scientists: James Watson, a tall, gangly biologist from America, and his fiery British colleague, Francis Crick. Beginning together at Cambridge, England, in 1951, the two attacked the mystery with relish.

Most biologists had assumed that genes were merely special protein molecules, much like the ones that make up hair or muscle tissue. But this simple solution was being disturbingly called into question by experiments which pointed toward a little-known molecule called DNA (deoxyribonucleic acid) as the possible carrier of heredity. If DNA was the answer, it would at last provide the Rosetta stone of life.

Beginning with an educated guess that the giant molecule was a twisting helix shape, Watson and Crick made models like Tinker Toys in their Cambridge laboratory, hoping to find just the right structure. Evenings were often spent in discussion at Crick's flat, where his French wife gave Watson a welcome respite from the "miserable English food." She didn't understand what their research was about—Crick had given up on her when she insisted that gravity extends only three miles into the sky—but already she looked forward to the day when he would be rich and famous, which to her meant being able to buy a car.

Meanwhile, across the Atlantic, Cal Tech's world-famous Linus Pauling was also working on the problem, increasing the pressure of competition daily. Crick and Watson, relying on luck as much as logic and cutting through academic rivalries and the bickering among fellow scientists, found more and more details falling into place.

When a simple mistake in calculations by Pauling set him suddenly behind in the race, Crick's supervisor, realizing the exciting possibility of beating the Americans to the momentous discovery, stepped up the research to full pace. New ideas were quickly tried and discarded until one morning, doodling on a piece of paper, Watson had a flash of insight.

The DNA molecule must be a pair of long, intertwined strands, he reasoned with growing enthusiasm, each a mirror image of the other. When the reproductive cells split, the strands must unwind and separate, with half the pair moving into the new cell. Thus, the genetic information is passed on to the next generation.

Crick was ecstatic when he heard the news and soon winged into their favorite pub, excitedly booming to everyone in sight, "We have discovered the secret of life!"

The Master Molecule

Indeed, they had discovered it. Publishing an understated nine-hundred-word announcement in the British science journal, *Nature*, Watson and Crick described the master molecule and revealed the secret of its awesome powers.

Inside every living cell, the genes along the chromosome strands are little more than DNA molecules. As Crick and Watson discovered, each of these molecules consists of two long strings of chemicals, twisted around each other like a braid.

Amazingly, the two strings are held together by only four simple chemicals, which also work in pairs. Every bit of genetic information is coded by the way these chemicals are arranged, like microscopic blueprints.

If you have green eyes, that is determined by a certain DNA molecule, while brown eyes are programmed by a different arrangement of chemicals along the same molecule. Whether your skin is white or black is the result of several molecules working together, as are body shape and other features. DNA is the chief executive of the cell in which it resides, giving chemical commands to control everything that keeps the cell alive and functioning.

The number of different gene combinations is so large that they form an almost limitless information storage system, like the memory bank of a giant computer. If the tightly coiled DNA strands inside a single human adult were unwound and stretched out straight, they would cover the distance to the moon half a million times. Yet when coiled, all the strands could fit inside a teaspoon. This powerful molecule is truly one of God's most amazing creations.

In the years following the DNA discovery, scientists have intensively studied chromosomes in an effort to map them, determining the location of the various genes and the specific order of the four chemicals which make up the genetic code. (The code was cracked in 1967.) Humans are believed to possess a hundred thousand or more genes (some say a million) on forty-six chromosomes, and only a thousand or so have yet been located.

The work of Crick and Watson triggered an explosive new era in biology, making it one of science's most active and productive fields. Advances in the past twenty-five years have been so great that the duo claim they could not even earn a PhD. now, much less a Nobel Prize, from such simple work as their momentous discovery of DNA.

In 1973, for example, an artificial DNA molecule was pieced together, while in 1978, a human gene was isolated for the first time, paving the way for its artificial creation. Able to design genes at will, scientists will soon have the awesome power to create totally new forms of life—plant, animal and human—with far-reaching implications for the future of our planet. The potential of our rapidly accelerating knowledge is nearly unlimited.

Artificial Life Next?

Will we reach the point of creating life itself, infusing into matter the vital force of which so many generations dreamed? Will faith in God plummet to new lows as man claims the ultimate power in the universe, proudly announcing himself the new master of his Father's limitless estate?

Pursuing the theory that life on this planet originated in the murky chemical soup which formed the primordial atmosphere, scientists have already conducted experiments in the creation of "life." Beginning in the 1950s, researchers artificially formed the basic ingredients of living creatures—amino acids and proto-viruses—by using heat or passing an electrical spark through such an atmosphere. Some of these molecules have been able to "reproduce," just as DNA splits off its own replica. In 1977, Japanese scientists claimed to have produced in this manner an artificial cell of very low complexity.

So intriguing are these experiments that Dr. Charles Price, upon his election as president of the American Chemical Society, publicly proposed that the creation of artificial life be made an American national goal, comparable to that of reaching the moon. With a major research effort, Price believes, small organisms could be made within twenty years, with intelligent artificial beings following in another century or so.[4]

The question dogging this research is, "What is life?" As most high school biology students know, it is almost impossible to say with certainty. Viruses, for example, have many characteristics of living organisms, yet they also act like nonliving molecules. Science fiction fans are well acquainted with biologically constructed space ships and organic master computers, able to grow new parts and heal themselves as animals do. But are these "living" beings?

If the day should come—and we can trust the scientists to continue working toward it—when an organism generally accepted as "living" is produced in the laboratory, how would that affect faith in the divine Creator of life? Would this be a victory for those who claim we are nothing more than collections of chemicals, worth only a few dollars on the open market?

Perhaps we should keep in mind that each great accomplishment of man, and every future triumph imaginable, is a crude duplication of the wonders God already has created. Far from being innovators, we explore a universe that predates any of us. We are like construction engineers who copy with reasonable accuracy a blueprint drawn long ago by Another, a master plan that holds countless pages we will never begin to understand.

If life were to be built out of nonliving matter, the "what is life?" dilemma might just be pushed further back, as we pursue the elusive spark into the inner mysteries of the atom and beyond. Since matter is merely another form of energy, as Einstein so brilliantly realized, the entire universe may be composed of nothing else than the energizing power of God's Spirit, with life only another form of His infinite working.

Assuming artificial life will be created someday, it might be an incredible new avenue for appreciating God's power and marveling at the wonders of His infinite plan for the universe. On the other hand, it could motivate mankind's highest expressions of arrogance, or lead him to despise himself as a cheap collection of molecules thrown together by accident.

The choice will be ours to make—one of many that clamor for our attention as staggering new powers leave the laboratory and touch our everyday lives. And nowhere are they more immediate than in the coming revolution in medicine.

5

Planet-Wide Plague

The baby's parents were a typical couple in their mid-thirties, with two young children at home. The hurried drive to the delivery room and the business-like bustling inside were almost routine for them, except that this baby was different. He suffered from Down's syndrome, or mongolism, and to make matters worse had a serious intestinal blockage. Without an immediate, relatively simple operation, the child would die.

His mother's immediate reaction was that she didn't want him. "It wouldn't be fair to the other children to raise them with a mongoloid," she declared emphatically. Arguments by the doctor that such a child, although possessing relatively low intelligence, would be perennially happy and able to perform simple tasks were to no avail. The husband stood by his wife's decision, insisting lamely that she knew more about these things than he did.

Honoring the parents' steadfast refusal to allow the lifesaving operation, the doctors made no attempt to save the baby or even to secure a court order overruling their decision. Instead, they wheeled him to a small side room at the hospital where, eleven days later, the tiny boy died from starvation, with the full knowledge of the anguished hospital staff. Had he been of normal intelligence, heaven and earth would have been moved to save him. As it was, all expressed a sigh of relief when the ordeal was over.[1]

We're All Victims

Everyone reacts differently to the news that his or her child has been struck by one of the more than two thousand genetic diseases that plague our planet. Few respond as the family in this recent Baltimore case did, but virtually all are shocked that such a thing could happen to *them*.

An estimated fifteen million Americans are ill today with inherited disorders, and a quarter of a million new sufferers are born each year. These diseases, the result of defects in genes and chromosomes transmitted by the parents, are said to account directly for an astonishing one-fourth of all hospitalizations and are strongly suspected to lurk in the background of many more illnesses—including cancer and heart disease.

We all have seen the posters of crippled children afflicted with such hereditary disorders as muscular dystrophy, but not all genetic illness appears in childhood. Huntington's disease, for example, the progressive degeneration of the nervous system that first became well known when folk singer Woody Guthrie was dying from its ravages, doesn't affect its victims until their late thirties. Glaucoma, another illness with a strong hereditary component, is also a late bloomer, destroying the vision of countless older people.

At this moment, each of us is carrying between three

and eight defects in our genes, "sleepers" that could show up at any time in our children. In addition, new defects occur randomly in every generation, more frequently in recent decades because of atomic radiation and chemical pollution. These genetic errors, symptoms of a distorted world that once knew perfection, can be transmitted to the next generation in several ways, with varying degrees of severity.

If a parent carries a faulty gene that is dominant, the risk is one out of two that a child will be affected by the disorder.[2] Called *dominant inheritance*, this accounts for nearly a thousand different illnesses, including the Huntington's disease that killed Woody Guthrie.

Much more common is the situation in which both parents appear normal but carry recessive harmful genes. These *recessively inherited* diseases are generally more severe than dominant ones, but the children run only a twenty-five-percent risk of receiving the defect their parents carry (and a fifty-percent chance of being carriers themselves). Only if a recessive disease carrier marries another carrier of the same disease do their children risk having the actual disease.

Among the nearly eight hundred known recessively inherited disorders are several that have received wide publicity. Cystic fibrosis is the most common, afflicting perhaps one in a thousand white children in America and carried recessively by about ten million of our citizens.

Another well-known recessive disease is phenylketonuria, or PKU. Because a child with PKU lacks the enzyme necessary to digest an amino acid found in most protein foods, the acid builds up in the body and can cause severe mental retardation.

Some genetic illnesses are called *sex-linked*, because they appear only in children of one sex, usually the male. In these cases, a boy has a fifty-percent chance of inheriting the disease, while a girl has the same chance of being a carrier.

One of the most well-known sex-linked diseases is hemophilia, the inability to properly form blood clots after being cut or scratched. It is popularly known as the disease of royalty, because Queen Victoria was a carrier whose daughters spread the illness throughout the royal families of Europe in the late nineteenth century. Eight of her grandchildren and six of her great-grandchildren were affected, including the son of Czar Nicholas II of Russia, whose illness led the parents to the strange monk Rasputin, who claimed he could cure the boy.

Still other diseases can be caused by the interaction of many genes, or by a malfunction involving an entire chromosome. Because genetic disease affects nearly everyone, researchers work constantly to isolate its causes. The National Institute of General Medical Sciences established seven centers in 1971 to study these disorders and now houses a collection of 190,000 frozen samples of diseased human cells, used by researchers around the world.

Calling for Counsel

As our knowledge of the cause and cure of hereditary disease grows, increasing emphasis is being placed on genetic counseling for those who run a high risk of having diseased children. Counseling centers are springing up around the country, with many tied to university medical centers.

Typical of the couples who visit a counseling center are Peggy and John Lipton, married for three years and now wondering whether they should start a family. Because Peggy's sister suffers from Down's syndrome, they were reluctant to consider having children. Through the persuasion of friends, they decided to seek professional advice.

The counselor began by taking a three-generation family history, looking for other genetic diseases in the couple's background that they were not immediately aware of. Peggy and John were amazed to realize how little they knew about their grandparents, but eventually the needed information was pieced together from relatives. The history looked clear, so the counselor focused on the mongolism problem.

"Because you're both young," he reassured, "your chances are very good. The incidence of Down's syndrome increases dramatically with the age of the parents. We'll make a simple chromosome analysis of Peggy to see if we can detect any problems."

A small blood sample was taken from Peggy's arm and grown for a few days in a culture medium. Then, the white cells were treated to give up their chromosomes, which were stained and photographed. Arranging the images in pairs by size, the technician concentrated on number 21 of the forty-six pairs, because at that point most Down's carriers and victims have an extra chromosome.

Much to her relief, Peggy was shown to have the normal number of chromosomes, so the chances of the couple's giving birth to a Down's child were remote. They departed happily, knowing they could look forward to a healthy family, with no more risk of genetic disease than anyone else.

Tests in the Womb

Most couples who seek genetic counseling today already have one child with a hereditary disease and are wondering about the next one, which all too many times is on the way. Here, the ending may not be as happy as with John and Peggy.

If the woman is pregnant, and counseling or testing indicates that the parents may have transmitted defective genes to their unborn child, the counselor might recommend tests on the baby in the womb to determine whether the defects appear in the growing child's body.

Major deformities can be discovered by several new techniques: ultrasound, a type of sonar that uses sound waves to give doctors a television picture of the fetus; radiography, the low-powered x-rays that reveal the structure of its skeleton; and fetoscopy, a fiber-optic flexible rod which allows the doctor to look directly into the womb through a small incision in the mother's abdomen and even take a blood sample from the fetus. But for most hereditary diseases, a chromosome and chemical analysis of the fetal tissue must be made, and the sample is usually obtained through *amniocentesis*.

First used less than fifteen years ago, amniocentesis has been heralded as the greatest breakthrough in medical genetics to date. For an average fee of three hundred dollars, the doctor carefully inserts a four-inch-long needle into the woman's anesthetized abdomen and withdraws up to an ounce of the clear yellow fluid that surrounds the fetus. By examining the small numbers of fetal cells floating in the fluid, technicians can now detect hundreds of genetic diseases.

Since entering public use in the early 1970s, the test has become so popular that the Federal government intends to promote it as a matter of official policy. Many doctors see amniocentesis soon becoming the standard medical procedure in the developed countries.

One reason why its popularity is growing among doctors is a recent New York Supreme Court ruling that held a doctor potentially liable for the lifetime care costs of a child born with Down's syndrome, because he failed to recommend amniocentesis for the thirty-seven-year-old mother. A lawyer in the case suggested that doctors may begin automatically prescribing the test as a protection against malpractice suits.[3]

Writer Jean Ashton describes what it was like for her to undergo amniocentesis:

> I saw poised above me a large syringe rapidly filling with a yellow liquid. It was removed while I watched and replaced by another. I seemed to be losing an enormous amount of fluid, but when I asked about this, the doctor answered me that it would be replaced within a few hours. . . .
>
> A Band-Aid was put on the puncture, my gown was pulled down, and that was that. I had felt no pain during the process, only a vague unpleasant ache, somewhere between a mild contraction and a menstrual cramp. It was only as I left that I thought about the possibility that the fetus might be injured or that the tests might otherwise fail.[4]

Risk, Error and Abortion
The question of injury or failure casts a shadow on

the bright picture supporters of amniocentesis paint. First, there is a small but definite risk (up to five percent) of accidentally injuring the unborn child with the needle or triggering a spontaneous abortion.

Then comes the problem of the accuracy and completeness of the test. In the case of Down's syndrome, reliability is close to one hundred percent, but other chromosome analyses can be inconclusive.

Tests may indicate the presence of a genetic difficulty, but doctors are often unable to predict what effect it will have on the baby. Laboratories fail to test for some diseases just because they lack the proper equipment, concentrating instead on the defects most easily testable. And in five to ten percent of the cases, a second fluid tap must be made because the first culture did not grow, greatly increasing the risk to the fetus (it can take a month or more for the results of each test to be determined) and the psychological hardship on the parents.

The most significant question hovering over amniocentesis is its close link with abortion. Dr. C. Everett Koop, famed surgeon at Philadelphia's Children's Hospital, calls it "a search and destroy mission."

Because our ability to treat genetic disease is so limited today, the option chosen by most parents when told their baby may be significantly abnormal is to have an abortion, usually well along in the pregnancy. Most doctors will not even perform amniocentesis unless, in the words of one expert, "the woman and her physician have made a 'firm decision' to 'interrupt the pregnancy' if a positive diagnosis of a particular disease is made."[5]

Concluded a major article on the subject in the prestigious *New England Journal of Medicine,* "The advent of prenatal diagnosis through amniocentesis represents the most important advance so far attained in the *prevention of births* of infants with irreparable genetic mental defects and fatal genetic disease."[6]

For those who accept abortion as a moral means of dealing with a diseased baby—and many who oppose abortion suddenly change their minds when faced with the prospect of giving birth to such a child—amniocentesis poses few problems. In fact, some doctors argue that it saves more fetal lives than it ends, because parents would terminate a pregnancy that runs a high risk of genetic disease unless convinced that their child is all right.

But what about those multiplied millions who consider abortion to be wrong? Aside from the tiny minority who wish to know in advance whether their child will be diseased so they can be psychologically prepared, amniocentesis has very little value for them. This fact may change as treatment for more hereditary diseases becomes available, but at present, amniocentesis most often provides information that medicine is unable to act upon, short of abortion.

This underlines the need for couples who suspect genetic disease in their family history to seek counseling *before* pregnancy, as Peggy and John Lipton did. There, they can learn whether or not they are carriers of a disease, and if so, what the chances are of transmitting it to their children. Only with full information can people exercise the wisdom God gave them and make responsible decisions about their future family.

Information, Please

The genetic counselor many times finds himself in a difficult moral position. Many genetic specialists hold the opinion, as expressed by one authority, that "parents have the right to know every bit of available information about the child they are expecting, regardless of any decisions they might make about terminating the pregnancy."[7] However, because of limited counseling resources today, the information is often given only to those who consider abortion an acceptable option.

Others believe that parents are told too much about their genes already, burdening them with needless fears. Because of the ever-widening gap between the knowledge genetic counselors have and that available to the worried parents sitting across the desk from them, couples are forced to rely more and more on the judgment of the professional.

In a study at St. Christopher's Hospital in Philadelphia, for example, only nineteen percent of the parents counseled on a certain genetic disease were able to write out even a basic understanding of it, despite several months of instruction by counselors, group discussions and the reading of simplified pamphlets.[8]

Compounding the problem is our limited knowledge about hereditary illnesses. Some people are *balanced carriers*, possessing the chromosomes for a disorder such as mental retardation but without the slightest sign of the illness.

If an adult had this defect, he could take an IQ test to determine whether or not he was retarded. But if the same results are drawn from chromosome analysis of a fetus, what can the counselor say? The child might be retarded or might not. Who should decide what action to take?

Many other hereditary diseases present similar difficulties in diagnosis. Chromosome defects may cause serious problems, but often they do not. Genetic disease ranges from mild to severe, given the same gene abnormality. Environment, too, plays a major role in determining the degrees of illness an individual will suffer.

Experience has shown that a hereditary chemical imbalance, such as PKU, may be present at birth but correct itself within a few months. Who is to know what the outcome will be in a particular case?

Genetic counseling is far from an exact science, yet some counselors are willing to quickly recommend abortion even if the predicted disease would not be a crippling one. The price of perfection seems to climb higher with each passing year.

Sickle-Cell Scare

Adding to the difficulties is the fearful social stigma that can accompany hereditary disease. While the possibility always exists that other family members could be afflicted as well, people are often reluctant to tell relatives about the problem, held back by feelings of guilt and shame. Such information can touch off a powder keg within the family, with fingers of blame being pointed in every direction and some resentfully wishing they had been left in ignorance.

Occasionally, genetic disease has become a political issue, spreading stigma through the society at large. One of the worst cases involved sickle-cell anemia. Although long understood as a hereditary illness that affects mostly black people, it burst into politics in the mid-1960s when research funding to find a cure was made by some black leaders into a benchmark of government commitment to civil rights.

Almost overnight, interest in the disorder flared into an obsession. In the excitement, the disease, touching fewer than fifty thousand sufferers, was regularly confused with the trait, carried without harm by more than two million American blacks. By 1972, Congress had passed the Sickle-Cell Control Act, providing money for states to launch screening programs, and the whole country jumped on the bandwagon.

Poorly planned information efforts convinced thousands of horrified black children, diagnosed as carriers of the *trait*, that they would die young because of the *disease*. And since no cure has been discovered, many experts questioned the value of the screening in the first place.

Soon, insurance companies were charging blacks higher rates, and six major airlines flatly refused to hire carriers of the trait, although they were perfectly healthy. Black leaders now reversed their position and cried that they were being stigmatized, with a few viewing the screening as racist genocide.

Amid the uproar, the purpose of the program—convincing carriers of the trait not to have children—was totally lost. Only recently has the furor died down, and several states have abandoned the screening altogether. This first attempt at mass counseling has left bitter confusion in its wake.

Following such public turmoil, many fear counseling almost as much as they fear genetic disease, reinforcing the idea that the ultimate family-planning decisions should be left with the parents themselves. And they can make responsible decisions only when they are informed of all the facts.

Reaching the Frontiers

Aside from abortion, what options are available to parents who discover that their unborn child is likely to have a genetic disease? At present, the choices are dishearteningly few, but rapid progress is being made on many fronts.

If the child suffers from cystic fibrosis, for example, medications and other treatments are now available that will allow many victims of this once-fatal disease to live long and nearly normal lives.

Babies born with phenylketonuria (PKU) can be effectively treated by means of a special diet which excludes milk products and certain other protein foods until age six.

Other advances, including microsurgery on babies, organ transplants and a host of electronic and mechanical devices, add to the hope that victims of many hereditary illnesses can have useful lives, despite their handicap.

But now, on the frontiers of medicine, new techniques are being developed which can effectively treat disease while the child is still in the womb. And further ahead are an astonishing array of powers that may push hereditary disease off our planet altogether.

6

Building Better Babies

At age three, Karen Spencer is a bright, alert child with chestnut brown hair and shining dark eyes. She laughs in delight as her mother bounces her on a knee, a happy youngster full of life.

Few would guess that she wasn't expected to be here at all. While still in her mother's womb, Karen was diagnosed as a victim of Rh disease—an incompatibility between the mother's blood and that of the child. As the unborn child's red blood cells are systematically destroyed, some babies are doomed before they take their first breath.

But Karen was saved by a risky, delicate operation. Knowing at twenty-four weeks that she would die unless something were done quickly, Dr. Robert Creasy of the University of California at San Francisco decided to attempt a blood transfusion in the womb. Because Karen was less than twelve inches long, with an abdomen smaller than a half dollar, it was an extremely difficult procedure.

Dr. Creasy first injected dye into the womb. As the tiny fetus swallowed it, the doctor was able to locate her intestines by x-ray. Then, carefully guiding the needle, he began trickling blood into the miniscule body.

The first transfusion required only two thimbles full of the life-giving fluid, with larger amounts on the second and third operations. Then, rather than risk additional transfusions, the doctor ordered a Caesarean delivery at thirty-four weeks. Following intensive care, Karen was sent home to live a normal, healthy life.[1]

Performed on occasion since 1972, this incredible operation has been surpassed by a more difficult one. Here, the womb is cut open and a tiny tube painstakingly inserted into one of the unborn baby's veins. Slowly, the child's blood is exchanged for fresh, Rh-compatible blood, and the womb is sealed up again.

As experience with this new science (called *fetology*) grows, doctors hope in the future to tackle heart and blood vessel defects and to be able to remove the excessive fluid build-ups which kill many babies. Eventually, they will attempt organ transplants in the womb. Sickle-cell anemia and Tay-Sachs disease are also strong candidates for treatment, because their chemistry is already known.

The Ultimate Cure

Exciting as recent advances are, the highest hopes are being held out for another development which promises eventually to bring an end to hereditary disease. This is the true genetic engineering technology of *gene surgery*. When developed, it will be the single greatest power of the new biology.

With gene surgery, scientists look for nothing less than editing the master tapes of life, snipping out defective genes from the microscopic chromosomes and replacing them with new ones.

Take, for example, a person suffering from galactosemia, a hereditary disease which prevents him from breaking down a simple sugar found in dairy products. Because he is missing the gene which tells the body to perform this function, the individual will suffer mental retardation and ultimately die unless kept on a strict milk-free diet.

Using a special technique, the genetic engineer has a solution. He carefully grows a virus with the missing gene and transplants it into a human cell. Like a microscopic mosquito, it drives a shaft deep inside and squirts its own genes into the cell's command center.

The master DNA is re-edited with new instructions, and the cell begins working properly. As the cells divide and grow, they pass on the corrected commands, replacing defective cells with normal ones and bringing about an internal cure for a once dreaded disease.

Although it may sound wildly futuristic, this very experiment was successfully performed in 1971 by three researchers at the National Institute of Mental Health in Bethesda, Maryland. Highly experimental and limited so far to isolated human cells in a laboratory dish—not inside the body—the technique shows great promise for future application to a hundred or so other genetic disorders.

Because of gene surgery's powerful potential, research is going full speed worldwide. The National Institute of General Medical Sciences alone awards well over a hundred million dollars a year for these studies.

The exact location of the gene responsible for our physical growth has already been discovered, as have the relatively few genes which produce the antibodies we use to guard against infection. Today, scientists can tell you that sickle-cell anemia is caused by a mistake on position 17 of the 438 sites that make up the beta-globin gene. Tomorrow, they may be able to perform gene surgery and correct the error.

Obstacles Along the Way

But significant problems lie ahead before the next major breakthrough in this field. As one writer summarizes the difficulties:

> Finding the right gene. Getting hold of it. Finding the vehicle to get it into the cells. Finding the right target cells. Getting the gene past the body's skillful defenses against invasion, and past the cell's own still-subtler defenses. All the while avoiding contaminants. At last making sure the new genetic material has the right conditions and neighbors within the cell in order to function. At every step insuring that no harm is done the patient.[2]

Still, greater obstacles than these have been overcome in genetics, and we certainly can expect them to fall before the researchers' advance as well.

More significant, perhaps, is the damage that might be done by their success. Our knowledge of genetics is so elementary that elimination of the so-called "bad genes" may prove to be a mistake. The sickle-cell anemia gene, for example, is believed to have arisen millennia ago in Africa as a defense against malaria. While causing disease today, the sickled red blood cell was at one time a lifesaver, and may be again.

Concludes Dr. Arthur Steinberg of the National Genetics Foundation, "We know very little about the value of a gene to a given race or to the species. We only know about its value to the individual carrying it, and then only in instances where the effect is severe. The gene responsible for cystic fibrosis may have been advantageous. . . at some time in the past. It may still be beneficial, for all we know."[3]

Given the complexity of the problem and the unknowns involved, some wonder whether research money might be better spent elsewhere. "Making a gene would be a very big, expensive effort," muses John O'Brien of the University of California at San Diego. "We have to ask whether, relative to such other needs as the conquest of cancer, it is worth the time and money."[4]

Ethical questions also intrude into the promises of gene surgery. A guiding principle of medical research is that of "informed consent," allowing the patient to freely give his approval to an experimental technique after being fully informed of the risks. With gene therapy, the patient cannot be sufficiently informed, because the side effects are unknown and potentially disastrous.

Michigan State University's Leroy Augenstein observes: "Suppose we were to find a virus which carried the necessary DNA for correcting diabetes and made all the boys very tall (good basketball teams) and raised their IQ's by fifteen points (no flunking out of school)." No one would object to this on moral grounds. "But suppose instead," he continues, "we were unlucky, and the virus contained not only a certain amount of DNA [to correct the problem], but additional DNA so that the group tested either went on to have defective children or developed schizophrenia. We would have a whole generation with extensive genetic changes *before we even knew they were in trouble.*"[5]

Furthermore, correction of a hereditary disease will allow the individual who carries it to pass the defect on to future generations, while he might otherwise have been sterile or died young. Successful gene therapy will increase the need for itself in each succeeding generation, while contributing to the overall decline in human genetic "quality."

The answers are not easy, and our ignorance vastly outweighs our understanding. Caution is an important part of the new biology, even with the most "obviously" beneficial techniques. When gambling with the building blocks of life we cannot afford to lose.

Screen That Gene!

Long before we are able to treat most hereditary illnesses, however, the cry is being heard for mass genetic screening of the population. Geneticists hope to reduce the "burden" people with these diseases are said to impose on society. "I have visions of a future genetic clinic in which a person will have not one but hundreds of his proteins analyzed completely in short order," remarks one biologist ruefully. "The results will be run through a computer, and a license to reproduce will then be issued on the basis of a passing grade with respect to his (or her) genes."[6]

Building Better Babies

Most newborns are screened for a few genetic diseases at the present time—PKU being the most common. Virtually no effort is made to keep the resulting information confidential, even if it might stigmatize the individual in later life. Today, only those diseases that can be tested easily and cheaply are subject to screening, but recent advances in computer chromosome analysis will expand the number quickly, reducing the time for a complete analysis from one day to minutes. Because the attitude of most lawmakers has been, "If you can screen cheaply enough, do it," mass screening for a large number of genetic defects may soon be upon us.

With one-fifth of the health care costs in the United States directly linked to hereditary diseases, economics alone make genetic foresight an attractive idea. Compared to the third of a million dollars required to institutionalize for life one child retarded by PKU, the $1.25 for testing him or her at birth, followed by preventive treatment, is a bargain.

Because ninety percent of couples wait until they have given birth to a diseased child before seeking genetic counseling, many experts believe everyone should be screened before marriage, just as a blood test is given prospective newlyweds today. Along with the marriage license, they could be offered a computer printout listing the odds of their producing children with a variety of genetic diseases.

While such testing would provide great benefits to society, it raises questions: Would the screening be mandatory or voluntary, and what happens if carriers of genetic defects decide to have children anyway?

Those who want mandatory screening argue that

75

only a compulsory system would be effective. Speaking at the 1979 National Symposium on Genetics and the Law, ethicist Joseph Fletcher said there are "more Typhoid Marys carrying genetic diseases than infectious diseases. We have the obligation to prevent their birth," he insisted, and if parents won't cooperate, then "coercive or compulsory control is justified."[7]

We have long prohibited marriages for genetic reasons, such as inbreeding within a family. Compulsory genetic screening takes this one step further, advocates explain. Says one expert, "If anyone thinks or has ever thought that religion, wealth or color are matters that may properly be taken into account when deciding whether or not a certain marriage is a suitable one, then let him not dare to suggest that the genetic welfare of human beings should not be given equal weight."[8]

Holes in the Screen

While granting this position, some question the advisability of mass screening on other grounds. Rejecting the comparison commonly made between screening and mass vaccination programs, bioethicist Marc Lappé insists that vaccination against contagious diseases is not a valid parallel. "The conditions being tested for are neither contagious nor, for the most part, susceptible to treatment at present," he points out.[9]

"Making the situation even more difficult," adds Dr. Laurence Karp of UCLA's Prenatal Diagnosis Center, "is the fact that the illnesses can be prevented only by preventing the 'victims' from ever being born." It is difficult to argue, he concludes, that the fetus would choose not to live instead of living an impaired life with a genetic disease.[10]

Adding to the dilemma is the fact that so many of the defects for which we are able to test are quite rare. Our ability to detect a disease, not necessarily its frequency or severity, seems to determine which illnesses are screened.

The tiny state of Rhode Island, for example, has been carefully screening each newborn baby for "maple syrup urine disease," which leads to mental retardation. How often does it show up? About once every ten years![11]

The best solution might be to make genetic screening widely available, but on a voluntary basis. It could begin with pilot programs to help eliminate large-scale mistakes such as those in the sickle-cell uproar. Screening and counseling services would be provided to all who want them, with the emphasis on counseling before marriage.

A program of public education would also begin, informing people about the available services and, just as important, what the test results mean. After the parents know the risks involved—if any—in having a family, the final decision would be left up to them. Those who refuse screening or fail to act on the knowledge they are given should be allowed to do so, or society may create more ominous problems than the birth of a few genetically diseased children, as we shall see in the next chapter.

On the other hand, notes Princeton University theologian Paul Ramsey, parents who discover they run a significant risk of bearing diseased children have a moral obligation to refrain from having them: "If the fact-situation disclosed by the science of genetics can prove that a given person cannot be the progenitor of healthy individuals, . . . then such a person's 'right to have children' becomes his duty not to do so." Just as the church, from the apostle Paul on, has promoted celibacy to the glory of God, he continues, it should advocate voluntary restrictions on reproduction for the sake of future generations.[12]

Recalling Defectives

And what of the diseased children who *are* born? What will be society's reaction when they slip through the screen for one reason or another? Sadly, they are often looked upon as mistakes who never should have been born, defective products that somehow missed the quality control section on the assembly line.

To ethicist Daniel Callahan, this relatively recent hardening of attitudes is a major step backward. Once, he recalls, the belief prevailed "that society as a whole should share the burden of caring for the defective, that the parents or family should not be left with the sole responsibility for [his] care and survival. . . ." This ethic affirmed "a joint sharing of responsibility by all and for all, so that those least equipped by nature or nurture to function would not be inevitable losers."[13]

But as the bonds of community and responsibility have weakened in modern society, the genetically diseased and deformed are being thought of in some quarters as just another pollution problem, second-class citizens who, in the words of Dr. Leon Kass, "need not have been, and who would not have been, if only someone had gotten to them in time."[14]

How many of us automatically think of a child with Down's syndrome as a "mongoloid," rather than a *person* who happens to suffer from mongolism? Such a child has only one extra chromosome among forty-six normal ones, yet we almost instinctively ignore the normal set and focus on the abnormality, as if the child were part of an alien species.

Regretfully noting the spread of this outlook, some are beginning to ask, "What price are we willing to pay for the perfect baby?" By our increasing use of genetic screening, amniocentesis and abortion on demand, the social and moral costs are steadily rising. How soon will they inflate to prohibitive heights?

A child born with hemophilia or muscular dystrophy when most of his fellow sufferers were destroyed in the womb may feel insecure at best, perhaps resentful at being brought into a world which shuns him as one who slipped through. The parents could well be looked down upon as irresponsible for not "taking care of the problem" in time.

Even normal children sense the shift in attitude. Dr. John Fletcher of the Institute of Society, Ethics and the Life Sciences studied twenty-five couples who asked for amniocentesis with the intention of aborting the fetus if defects were found. In each case, he observed a subtle change in the relationship between the parents and their children. As one parent put it, "What is Johnny going to think about us now? Is he going to wonder, 'What would Mommy and Daddy have done if something had been wrong with *me?*' "[15]

To Dr. Kass, the principle, "Defectives should not be born," is one that has no limits, like a warm bath that heats up so imperceptibly you don't know when to scream.

Drawing the Line

Where do we draw the line between defective and normal? Mental retardation, like other hereditary problems, can be mild or severe, but how severe is severe? When screening becomes standard medical practice, theologian Paul Ramsey warns, the criteria for normality are bound to be upgraded, while the level of care for abnormals is at the same time downgraded.[16]

There may even come a time when infanticide is reintroduced, as advocated by a few today. Writes Gerald Leach in his book, *The Biocrats*, "By infanticide I mean the killing of a baby because its parents make a hard-thought-out decision that to keep it alive involves a greater sacrifice for all, and is more offensive, than killing it. . . . Perhaps we are going to have to make a chronologically minute, but emotionally vast, shift in the *dating* of our acceptance of a baby into society. . . . It may be that two or three days' or weeks' 'probationary life' should be accepted as a period during which doctors could check for defects and parents could decide whether or not they wanted to keep and rear a damaged baby."[17]

Impossible? Society made nearly as radical a turnabout in the past decade, shifting from a strongly anti-abortion position to its full acceptance, using the same type of dating shift to declare a child human later in its development. Given the common notion today of interchangeable fetuses—abort the "bad" ones until a "good" one comes along—it is not inconceivable that our pursuit of quality might someday lead in the horrifying direction of infanticide.

Before we reach such a low in human compassion,

we would do well to heed the warning thundered long
ago through the prophet Isaiah:

"Woe to him who strives with his Maker,
      an earthen vessel with the potter! . . .
   Woe to him who says to a father, 'What are you
      begetting?'
      or to a woman, 'With what are you in travail?' "
   Thus says the Lord,
      the Holy One of Israel, and his Maker:
   "Will you question me about my children,
      or command me concerning the work of my
      hands?"[18]

Yet some, ever ready to take the future into their
hands and become as God, are now setting human
perfection as their absolute goal. Adapting present
knowledge and anticipating future developments,
these dabblers with destiny have nothing less in mind
than the total re-engineering of mankind. It is a
frightful thought, but hardly a new one.

Fellow Mutants, Arise!

She thought her name was Irene de Fouw.

About all she knew of her background was that her parents had been war resisters, dead in a German concentration camp. That left her one of the countless thousands of orphans who flooded postwar France, with only a few shreds of information about their heritage on which to build a fragile identity.

One bright spring morning in her fourteenth year, Irene was tidying up the attic for her foster mother when she made a startling discovery. Reading a bundle of letters about herself, she trembled as she found contradictions which cast doubt on all she had been told about her parents. Questions directed to her stepmother were fended off with a harsh rebuff, as if some dark mystery existed which was best left hidden.

But Irene had to know. Four years later, she left home and began the long and frustrating quest for her identity. It took nearly a decade, but finally she found the incredible answer.

Her real name was *Ingrid* de Fouw, born July 31, 1944, at a Nazi maternity home in Lamorlaye, France. She had been bred as part of Germany's efforts to produce a master race—the *Lebensborn* organization, which united fathers from Heinrich Himmler's infamous SS with racially pure mothers to build an advance guard for the coming world empire.

Like the thousands of others born in these baby factories, Ingrid has no idea who her parents are. Today, she continues the hopeless search to find them.

Sounding like the wildest excesses of fiction, this monstrous experiment in the breeding of superior humans actually took place, and its victims—most in their thirties—still haunt Europe. Red Cross posters dot Germany with pictures of pleasant-looking young men and women staring out below the plaintive heading, "Who can tell us what our names are? Where we came from? Where we were born?"

After three years of intensive research, French journalist Marc Hillel and his wife Clarissa Henry recorded the astounding history of the Lebensborn (meaning "fountain of life") organization in their film and book, *Of Pure Blood*.[1]

However crude the Nazi racial beliefs and unscientific their methods, their goal of breeding a purified human race finds a chilling echo in the ideas of some prominent geneticists today. A look at the Nazi experiments and the walking wounded they produced is a fitting introduction to the schemes of the new genetics for perfecting humanity.

A Fair-Haired Empire

Long before the Nazi rise to power, SS Reichsfuehrer Himmler dreamed of a racially pure empire in which births would be regulated by strict selection. His goal was to populate Germany with 120 million fair-haired

Nordics by 1980 at the latest.

Once the swastika flew over Germany, he lost little time in organizing the Lebensborn Society, opening its first center in 1935. There, unwed mothers of "racially valuable" stock were given free maternity care, and soon teenage volunteers also arrived. Stirred by intensive propaganda appeals to "bear a child for the Fuehrer," the naive girls were impregnated by tall, blond SS men, who could document Aryan ancestry at least back to the eighteenth century.

Lavishly financed with money confiscated from Jews and other enemies of the state, the Lebensborn organization mushroomed. Hospitals, clinics and maternity homes in conquered countries were seized and turned over to it, their patients often packed off to death camps. By 1942, Himmler was finalizing plans to build a huge birth center which would process 400,000 unwed mothers a year. The new technique of human artificial insemination also aroused great interest, but the war sapped the research effort necessary to perfect it.

The principal opponent to these schemes was the Christian church, especially Catholicism, considered "the enemy of fertility" by the Nazis because of its teachings on marriage. In response, Lebensborn nurses signed a declaration rejecting Christianity in favor of the new religion of pure blood, and christenings were replaced by a bizarre Nazi name-giving ceremony.

Abnormal children born in the centers—and they had their share of mongoloids, retarded children, paralytics and other diseased babies who were not the *Edelprodukt* (top-quality goods) that Nazi racial genetics promised—were summarily sent to a psychiatric clinic near Potsdam. There, they were "disinfected," their tiny bodies dissected in the name of scientific research.

Imported Children

As the war dragged on and the once-invincible German military machine was decimated on the battlefield, the supply of Aryans fell alarmingly. Now, the Lebensborn organization ordered the wholesale kidnapping of suitable children—in the street, at home, at school—from the occupied countries. More than 200,000 youths were wrenched from their parents in Poland alone, with tens of thousands more in Norway, France, Holland and other countries, all carried off to be raised as Germans.

Given new names and false birth certificates to cut every tie with home, these helpless victims were saturated with Nazi propaganda. They were to be abandoned as the Reich crumbled, left for the conquering armies to house and feed.

Few to date have ever found their parents again, and are condemned to carry permanent psychological scars. Far from a superior race, they are today the misfits of German society, without ancestry, without a true home.

At Nuremberg, leaders of the Lebensborn movement were put on trial, but were merely found guilty of belonging to the SS. The Lebensborn society, the court decided, was a welfare organization and had nothing to do with the mass kidnappings. Its chief administrator is still alive, claiming that "everything we did was done in the name of charity, in the name of mercy."[2]

Asked whether the children rescued from Lebensborn

homes acted like part of a super-race, the German medical officer put in charge by the Americans after the war scoffed, "Most certainly not. The number of backward children among them was far above average, about fifteen percent. . . . These children did not know what tenderness was. They were . . . frightened of any grown-up who approached them."[3]

Geniuses and Giants

Surprisingly, the idea of improving the human race through selective breeding is not a new one. Leonardo da Vinci's half brother, for example, scoured the countryside, looking for a wife who resembled the genius's mother. Finding such a girl, he had a son, Pierino, by her and raised him to be an artist. The highly intelligent boy showed a remarkable aptitude for design and produced works later attributed mistakenly to Michelangelo. Unfortunately, he died of fever at age twenty-three and never fulfulled his promise as a second Leonardo.[4]

Frederick Wilhelm of Prussia was even less successful in his attempts. Trying to breed a race of giants for his personal guard, he was disappointed to discover that many of the fathers were tall because of glandular problems, not heredity. As a result, they had difficulty producing healthy children of any height, let alone giants.[5]

In the mid-nineteenth century, influenced by the Enlightenment and its unshakable belief that man can be perfected, proposals to improve the quality of the human race abounded. The publication of Darwin's theory of evolution sparked the flames, and Darwin's brilliant cousin, Francis Galton, fanned them into a conflagration that soon swept the Western world.

Coining the term *eugenics* (from the Greek for "well-born") for the breeding of superior human stock, Galton preached tirelessly on its behalf, pointing to successes in agriculture and livestock breeding as proof that the species could be improved. He was inspired by the popular belief that heredity is carried in the blood, and put his appeals in blatantly racist and class-oriented terms. The pedigreed "bluebloods" were encouraged to have more children to prevent their being engulfed by the larger numbers among inferior strains.

Galton played heavily on the snobbishness of the upper classes, among whom science and reason were fashionable cure-alls in that golden age of optimism. His great success was multiplied by the Social Darwinists, who gleefully supported the worst abuses of unrestrained capitalism and racism in America. Rallying around the cry of "survival of the fittest," they built such a following that Justice Oliver Wendell Holmes was forced to remind his Supreme Court colleagues that their theories were not part of the Constitution.[6]

Social Darwinism promoted unlimited competition as the key to progress. Those who fell by the wayside, such as the poor, were unfit to compete anyway and should be left to die out. Since American business and intellectual leaders came mostly from old Anglo-Saxon families, it was easy to convince them on "scientific" grounds that the Nordics were nature's fittest race and should be given preference over inferior stocks.

America, Home of the Pure

German, British, Russian and other eugenics societies sprang up to advance the cause of a purer race, but none of them met the success found in America, where progress has long been a national idol. "I wish very much that the wrong people could be prevented entirely from breeding," said President Theodore Roosevelt. "Criminals should be sterilized and feeble-minded persons forbidden to leave offspring behind them. . . . The emphasis should be laid on getting desirable people to breed."[7]

Until World War I, nearly half the scientists involved in the new field of genetics were enthusiastic members of the eugenics movement. Their fears about the decline of the human species were matched by their belief that science could take care of all mankind's ills. Research in eugenics and genetics was commonly carried out by the same scientists, with other members of the movement prominently including educators, clergymen, scholars and doctors.

Basing their theories on mistaken ideas—especially the belief that single genes caused such "traits" as alcoholism, criminality and laziness—the eugenicists drew elaborate charts of genetic superiority, built around the most blatant ethnic stereotypes. One chart, for example, listed the races in order of perfection, ranging from the Negro at the bottom up to the exalted Caucasians, who were in turn ranked in ascending order as Mediterraneans, Alpines and Nordics.

Famed New York eugenicist Madison Grant disparagingly described Jews as self-interested and Catholics as ignorant and destitute. Negroes were painted as "willing followers who ask only to obey," while he considered Indians to be so cruel they could "scarcely be regarded as human beings." Only the Nordics escaped his broad brush of prejudice, a race known for "leadership, courage, loyalty, unity, self-sacrifice and devotion to an ideal."[8]

Becoming uneasy with the extremes of the movement and discovering that its scientific basis was faulty, geneticists began pulling out during World War I, but the damage had already been done. By 1915, eugenics was a common part of the social scene in America, the subject of popular books, lectures, magazine articles and sermons. Universities taught courses on it, notables such as Alexander Graham Bell spoke on its behalf, and even the fledgling Boy Scouts aimed to rear the "eugenic new man."[9]

Search for Superiority

Concluded University of Wisconsin President Charles Van Hise in 1914, "We know enough about eugenics so that if the knowledge were applied, the defective classes would disappear within a generation."[10] But after World War I, a flood tide of immigrants, mostly from Southern Europe, inundated American shores, and this hope rapidly faded. A wave of anti-immigration hysteria swept the nation.

For the eugenicists and their theories of Nordic superiority, this outpouring of emotion was just the boost they needed. They began noisily campaigning for laws which would restrict the immigration of "inferior races." Congressional hearings were held, dominated by discussions of race and superiority, and the great melting-pot ideal drowned in the deluge. In the end, the Immigration Restriction Act of 1924 passed by a broad, bipartisan majority, placing into law a brazen discrimination against non-Anglo-Saxons that would remain on the books until 1965.

Summarizing the position of the Coolidge Administration during the debate, Secretary of Labor James J. Davis said, "America has always prided itself upon having for its basic stock the so-called Nordic race. . . . We should ban from our shores all races which are not naturalizable under the law of the land and all individuals. . . who constitute a menace to our civilization."[11]

During the same years, agitation grew for passage of sterilization laws to prevent criminals, the mentally retarded, epileptics and others from passing on their "disease." Upheld by the Supreme Court in 1927, these statutes were eventually written in thirty states. They led to tens of thousands of forced sterilizations, opposed with any vigor solely by the Catholic church.

Margaret Sanger, heroine of today's feminists for her pioneering efforts with birth control programs, wholeheartedly supported the effort. "It is a curious but neglected fact," she said, "that the very types which in all kindness should be obliterated from the human stock have been permitted to reproduce themselves and to perpetuate their group. . . . Sterilization is the solution."[12]

Decline and Fall

But as the eugenics movement peaked in the 1920s, the seeds of its decline were already being sown. Increased understanding of genetics and the role of environment in behavior cast grave doubts on the eugenicists' theories. Some of the most publicized pedigree searches were proven to be frauds, little more than the imagination of amateur genealogists.

A particularly embarrassing setback came to light after the First World War. When the IQ test used by eugenicists to prove racial inferiority was given to U.S. Army draftees, it rated half of them as feebleminded. Worse from the eugenics point of view, the test gave Northern blacks consistently higher scores than Southern whites!

The death blows to the early eugenics movement fell in the 1930s. Reports of Nazi racial policies (initially applauded by American geneticists) finally spurred horrified scientists to actively oppose the movement. The Depression nailed the coffin shut, as Nordics were forced to stand in breadlines alongside the despised Jews and blacks.

One poignant example of the damage done by eugenics can be found among the "poor white trash" of the South, a constant source of shame for the eugenicists. They had been explained away as a sub-race with inferior blood, inheriting the disease of chronic poverty as part of their genetic makeup. Although medical studies had established that their "laziness, shiftlessness and ignorance" was due to pellagra, an often-fatal disease caused by malnutrition, eugenicists heatedly denounced this explanation, and treatment was rarely given.

During the Depression, however, government food programs began nationwide, and the disease virtually disappeared. A change in diet—not genes—had quickly solved the problem after millions of the poor had needlessly died, victims of the pseudo-scientific prejudice that swayed a nation.[13]

Fellow Mutants, Arise!

As eugenics was fading in America, it laid the groundwork for a new social order in Nazi Germany, built partly on the information obtained from American sterilization programs. Within months of his rise to power, Hitler passed a sterilization law which led to a quarter of a million operations. Euthanasia ("mercy" killing) of the feeble and infirm was legalized in 1939, and its tens of thousands of victims served as practice for the mass extermination camps to follow. Eventually, multiplied millions were murdered for reasons of racial purity, feeding the beast of eugenics its inevitable diet of human agony.

The New Eugenics

After the Nazi atrocities and the horrors of racism were exposed, we might expect the subject of purifying the human species to be given an uneasy place on the back shelf of history, lived down along with the worst that man has produced. Surprisingly, eugenics is now enjoying a comeback, based on updated scientific knowledge but playing a familiar theme.

As before, it begins with doomsday prophecies about the decline of the human species. This time, however, the decline is described in terms of the growing genetic load in the gene pool (i.e., a larger proportion of genes which produce disease and disabilities among the population) instead of race and class deterioration.

The modern-day eugenicists insist that if something isn't done quickly, we will become a society of cripples and sickly retards. One geneticist recently warned that within a few generations we will see one child in every ten with serious defects unless steps are taken soon to halt the decline.[14]

Opponents are not so sure that humanity is deteriorating. They hold that the diversity of genes is one of our greatest assets, allowing a variety of responses to a changing environment. Because so many traits involve more than one gene, it would be very difficult to eliminate the majority of genetic diseases—and we might wipe out the key to a future generation's survival in the process.

Furthermore, facts about genetic decay are extremely sketchy, because the subject has only recently been under study. In a test of intelligence among Scottish schoolchildren, one group born fifteen years after the other, the results showed a significant *increase* in IQ in the second group, rather than the decline eugenicists predicted.[15] Much more research lies ahead before we begin fooling with our future.

As with the earlier eugenics movement, debate over the decay of the human race is quickly giving way to grand social programs for creating model citizens and a golden age of society—free of crime, turmoil and the burdens of overpopulation. Today, the old argument of heredity against environment is back in full force: Are misfits born or made?

Since the discrediting of eugenics forty years ago, the majority has clung to the belief that environment controls behavior, leading to such fading hopes as rehabilitation of prisoners. Now, the pendulum is swinging back to the side of heredity, and the proponents of human perfection are once again ready to rise.

8

Quest for Perfection

Will parents of the next century need licenses in order to have children? That is the shocking plan advocated by growing numbers of geneticists as part of a far-reaching program to improve the human race.

According to one scheme, each girl (or boy) is sent to the doctor at puberty and implanted with a "contraceptive time capsule," perhaps inserted under the skin. (Development of these capsules is already well underway.) When the youth grows up and wants to start a family, he or she obtains a license from the government and returns to the doctor for temporary removal of the capsule.

Not only would the system cut down the number of abortions, but it would give the sexual revolution its greatest boost ever. And in the spirit of free enterprise, additional licenses could be sold on the New York Stock Exchange to those who desire larger families. The proceeds would go to the patriotic citizens who have fewer children than the permitted maximum.

Who would be granted licenses? One idea is to require graduation from a community college course in child rearing as a prerequisite. Another allows a couple two children, but demands a detailed physical and mental examination of the first two before permitting any more.

Needless to say, carriers of genetic defects need not apply for licenses at all.

The authors of the licensing plan are convinced that Americans will accept a system like this within a generation. Comments ethicist Joseph Fletcher on these proposals, "We will soon see how absurd it is that we have to be issued drivers' licenses to operate an automobile, while we are free to go on producing children of any old kind we want or happen to conceive."[1]

Pursuing Quality

As the eugenicists plan their strategy for improving the species, they see population control as a first step toward quality control. "We do not have many decades before we will be forced to stabilize the size of the population," observes Dr. Bernard Davis of Harvard, "probably by accepting external restrictions on our freedom to procreate; and the step from quantity control to selective control may then not seem a large one."[2]

Controversial studies of human "quality" are taking place right now, adding fuel to the eugenics fire. The man most prominently associated with this research is Arthur Jensen, professor of educational psychology at the University of California.

He holds the opinion that, for intelligence at least, heredity counts for eighty percent of the differences between people, and environment only twenty. (Most geneticists say they are about equal.) Furthermore, he insists that the races have different intellectual abilities, with blacks lowest and becoming ever lower compared to whites.

This thought has been picked up by other leading scientists and educators and is considered serious enough for the Federal government to fund a $1.7 million study on the relationship between intelligence and race. In the grant application, it was ominously pointed out that "these data will serve as a basis for future decisions . . . about population control which will have to be made at Government level."[3]

Another eugenicist, the Nobel Prize-winning co-inventor of the transistor, William Shockley, feels no need to wait for further study. He came up with the idea of paying people $1,000 for each IQ point below 100 if they would consent to be sterilized. Since Jensen insists that blacks score an average of fifteen points below whites on intelligence tests, that would amount to a cash grant averaging $15,000 to every black American for not reproducing.[4]

Shockley insists his motives are humanitarian: "If, in the United States, our nobly intended welfare programs are indeed encouraging the least effective elements of the blacks to have the most children, then a destiny of genetic enslavement of blacks may well ensue."[5] While presented in milder terms than the earlier eugenicists used, such proposals sound fearfully familiar.

The Criminal Chromosome

It doesn't take much to stir up public demand for eugenic controls. A decade ago, several studies became public which indicated that a male born with an extra Y (male) chromosome was more likely to end up in prison or a mental institution than one with the normal number. Simultaneously came the mistaken report that mass murderer Richard Speck was one of these "XYY" types.

Although refuted by other studies, this so-called genetic basis for criminal behavior gave sensational new ammunition to believers in human perfection. Several prominent officials called for registration of all XYY males at birth and a close watch on them in subsequent years.

Hospitals in a few large cities began routinely screening for the "criminal chromosome," and Dr. Bentley Glass, former president of the American Association for the Advancement of Science, wrote in *Science* magazine that he looked forward to the day when amniocentesis and state-regulated abortion would rid the world of the XYY type.[6]

The furor has since died down, as further research disproved the XYY theory, and many scientists now scoff at the notion that heredity influences any kind of behavior, criminal or otherwise. Yet whether the intent is to stop a supposed decline in the species or engineer a super-race, eugenicists continue to plow ahead, despite all obstacles.

Julian Huxley, noted biologist and brother of *Brave New World* author Aldous Huxley, concluded: "It is clear that for any major advance in national and international efficiency we cannot depend on haphazard tinkering with social or political systems . . . , but must rely increasingly on raising the genetic level of man's intellectual and practical abilities. . . . To be effective, such non-natural [human] selection must be conscious, purposeful and planned."[7]

Already, precedents have been set for a response to

Huxley's call. Denmark, hardly an uncivilized country, requires sterilization of any woman with an IQ less than 75, and there seems to be no legal reason why similar action could not someday occur in the United States.[8]

In a 1965 Supreme Court case, Justice Arthur Goldberg wrote that mandatory sterilization would be unconstitutional *unless compelling social reasons could be shown.*[9] With rising population and growing unwillingness to accept genetic disease, such reasons could well be imagined. After all, the government is saddled with increasing costs of schools, medical programs, crime prevention and welfare. Why shouldn't it control the way that money is spent?

Most likely, a program like this would begin with the disadvantaged in society—with population control first—and then move up the social ladder until all were included. A decade ago, the Department of Health, Education and Welfare gave welfare case workers permission to urge contraception on poor families. At the time, the Secretary suggested that "disincentives" be offered in the future to discourage welfare parents from having more than two children.[10] Could this be a first step toward mass licensing?

The Coming of Superman

Without waiting for licensing programs, eugenicists eagerly plan for the day when the human race can be genetically re-engineered, breeding "supermen" to outperform the people of today in every area of achievement. From crude methods such as artificial insemination by "superior" donors to sophisticated techniques of gene surgery, they envision a new era for mankind through the magic of genetics.

Biomedicine, said Dr. Charles Frankel of Columbia University, claims to be bringing us to an age "when man will be able to say of himself, meaning it entirely, that at last he is his own greatest creation, and has got the weight of that other Creation off his back."[11]

Finding genetic manipulation a much more promising way to improve mankind than the social reforms which have failed so miserably, the eugenicists have a variety of schemes in mind. Quality of parenthood might be improved beyond the licensing system by converting sperm banks into "seed markets," selling a Who's Who list of sperm and eggs for a premium price. If you are going to buy a donor seed in order to have a baby, they ask, why not get the best? Selecting from a glossy, full-color catalog which describes features of donors and discreetly lists prices, the proud parents will know exactly what they are buying. Faulty products will of course result in an immediate refund.

For old-fashioned parents who hold on to the romantic notion of raising a baby of their own, multiple parenthood might be the answer. The couple would contribute their own genes, with genes from one or more superior donors added to bring the child up to the desired quality.

Since intelligence has such an overwhelming grip on Western culture as *the* most desired trait, it will surely be one of the first receiving the eugenicists' attention. Experiments are being done today on growth modification of the brain; a simple growth hormone injected into a rat fetus has resulted in a seventy-five-percent increase in brain size, with an equivalent rise in learning ability. With the advent of genetic engineering, much more sophisticated techniques will certainly be available, perhaps to double the IQ of every newborn baby.

Green People?

When discussion turns to modification of the human body, the ideas pour out at an astonishing rate. The British *Science Journal* asked its scientist readers for redesign suggestions, and their proposals were mind-boggling.[12]

Design man with gills so cities could be built underwater, some readers said. Enlarge the appendix to enable humans to eat grass like a cow and digest the cellulose, suggested others. Separate the esophagus from the trachea so we won't choke on our food; give everyone photographic memories; enable our heads to swivel 180°, or put our eyes on stalks so we can see in all directions.

Not to be outdone, *Sciences*, the publication of the New York Academy of Sciences, made a similar survey. "Man should be made smaller in size," said Nobel Prize-winner Charles Townes, reasoning that modern technology no longer requires large bodies.

"Learn the art of photosynthesis," London University's Dominic Recaldin proposed, allowing us to convert sunlight into food as a plant does. An added benefit, he noted, would be the end of the race problem: everyone would be green!

Another startling thought is to eliminate breast cancer by designing women without breasts. Infants could be fed anyway, because milk production is not related to breast size. Perhaps this would be the ultimate in women's liberation, allowing freedom from an encumbrance in factory jobs and combat duty, while at the same time removing part of the sex object mystique denounced by the feminists.

Chimeras: Built to Order

One of the most radical ideas for redesigning humans is the combining of genes from plants and animals with those of man to form a "chimera." Human and mouse cells were successfully fused into one as far back as 1967, and since then human cells have been combined with those of both tobacco and carrots as part of ongoing cancer research. Who can believe that such grafting might not become commonplace with humans in a few decades?

Renowned geneticist Dr. J.B.S. Haldane proposed a chimera astronaut adapted to the special environment of space as an example of what might be accomplished.[13] His creation includes the agility of an ape and a tail to aid in swinging around a zero-gravity space ship. He leaves no place for legs, which would only get in the way.

Perhaps the astronaut could also be made germ-free inside and out to avoid contaminating other planets. Of course, he would loathe contact with normal humans, because we would smell to him like foul barnyard animals.

As our high-technology society becomes more specialized, the design of human and sub-human chimeras will find a ready market in industry. A few companies are already screening workers' chromosomes for traits that might cause problems with their jobs, such as Dow Chemical's screening for susceptibility to chemically induced diseases. Psychological testing, widespread at present, might be replaced by gene analysis to find workers who "fit" in a particular industry, eventually leading to specialized workers designed for specific jobs.

While higher intelligence is a common goal, lower brain power might also be desirable for more monotonous jobs. *Time* magazine reported that some restaurant owners actively seek out the mentally retarded today because they are the only ones who take some pride in mopping floors and washing dishes.[14] This trend might be expanded with sub-human chimeras, costing much less than the servant robots we all have seen in science fiction movies.

At what point would a chimera be declared human? Two lawyers gave some thought to the subject and found it extremely complex. Should personhood be granted to a being who is one-hundredth human? A tenth? More than half? Can a soul be divided into fractions?

As pressing is the question of what those designed to be astronauts, welders or super-secretaries might think of their pre-ordained role. How could someone engineered to be a mechanic ever hope to find a job if he wanted to be a chemist, for example? Schools are widely criticized for placing children into a "tracking" system, some for college, others for trade occupations. Genetic tracking would form an almost unbreakable slavery.

Resistance and Risk

As genetic engineering knowledge increases, making widely available the ability to multiply intelligence along with controlling height and weight, parents who object on moral or religious grounds to manipulating their children might be looked upon as cold and uncaring. Most likely, they would be refused licenses for children if such a system were in operation.

Suppose the average IQ among children suddenly shot up to 200. What would be the fate of those youngsters whose parents refused the modification? They could well belong to a "genetic ghetto," ridiculed by the other children as morons and thrown into the most menial jobs in adulthood.

Early Christians were branded traitors because they would not conform to requirements by the government which they considered wrong. Could eugenics bring a future wave of persecution against those who hold religious principles to be higher than social improvement schemes? How many parents could withstand the social pressure to join in building the new super-race?

A host of other serious questions are raised by the new eugenics, which has a potential for redesigning humanity—and for disaster—that the old eugenicists never dreamed of. The effort to improve the quality of the human race would be incredibly complex and full of failures along the way. Knowing so little about genetics and the interlocking systems among genes, we may be sorry if we tamper with them.

Licensing superior parents would probably yield disappointing results. From animal breeding, we know that the response to intense selection for a given character trait begins rapidly, but then slows down over a few generations and at a certain level stops completely. In addition, selective breeding always exacts a price. You can breed pigs for good bacon yield, but they end up stupid, docile and unable to walk more than a few yards. Breeding for all-out improvement is just impossible.

Applying sophisticated gene surgery techniques might be more effective, but the potential for disaster is infinitely greater. Most genes operate in groups; modifying a gene we know about in order to raise intelligence or increase strength may affect some other trait in a way we don't anticipate. Worse, the harmful effects might not show up until several generations had passed, after the catastrophe had spread throughout the population.

A Veiled Future

Here lies the heart of the problem with eugenics: Not being God, how can we expect to plan effectively for an unknown future? As an example of our failures in forecasting, take the book published in 1962 by the influential Kiplinger magazine that attempted to show what new scientific wonders would fill our lives in 1975.[15]

Basing the book solidly on developments well underway at the time it was written, the writer still missed on more than half the predictions and failed to mention the really widespread innovations such as miniature computers and microwave ovens. The automobiles of the 1970s looked like 1959 Cadillacs with a futuristic plastic dome and rakish fins tacked on (turbine powered, of course). Experts predicted atomic-radiated food that would keep indefinitely without refrigeration. Our food probably *is* radiated today, but not deliberately!

If we cannot even project developments over a fifteen-year span, isn't it a mistake to attempt a re-engineering of future generations for a world only dimly imagined?

We laugh at the "scientific" ideas firmly held by the best minds half a century ago, such as the blood and race theories of heredity or the existence of lush green canals on Mars. Many of today's beliefs will sound just as ridiculous in the next century.

How then can we presume to think about redesigning man, even for an "obvious" advantage like higher intelligence? We already have many more PhD.s than jobs for them. Perhaps a skilled mechanic will be the person most in demand in a future generation—or perhaps not.

Notes theologian Paul Ramsey: "Mankind has not evidenced much wisdom in the control and redirection of his environment. It would seem unreasonable to believe that by adding to his environmental follies one or another of these grand designs for reconstructing himself, man would then show a sudden increase in wisdom."[16]

Since every generation seems to blame the preceding one for the faults of the world they inherit, what would they say to being physically and mentally engineered by their failure-prone ancestors? As one philosopher has pointed out, to navigate by a landmark tied to your own ship's bow is to invite shipwreck.

Eugenicists assume that smarter people would be better people and, furthermore, that genes can do the job of raising intelligence. Neither of these ideas has been proven. Smarter people also make better criminals or self-centered, fast-talking con artists.

And what about parents trying to raise children with IQs double their own? Imagine the difficulty of prying the truth out of the neighborhood gang of little Einsteins when they are trying to get away with something.

As in all the schemes for improving mankind, from Plato's *Republic* to B.F. Skinner's *Walden Two*, the eugenicists propose to build up society (or the species) at the expense of the individual. Personal human life is less respected than a dream-like ideal.

In fact, however, parents are responsible for *their* children, not the race. Doctors, too, are sworn to help their patients to the best of their ability—man, woman and unborn child alike—rather than serving the needs of the species or some grand improvement idea. Despite the sweeping genetic decline predicted by eugenicists and the tremendous social burdens they foresee, the question really comes down to the individual level.

As bioethicist Marc Lappé points out, "The notion of a genetic 'burden' imposed on society by individuals carrying [harmful] genes is a misleading concept: the 'burden' . . . is borne by families, not society."[17] Thus, he concludes, families and their doctors are the ones to make the agonizing decisions, in the context of a particular God-given life, not a society bent on self-evolution.

Who Is the Ideal?

In any case, people cannot agree on what the "ideal" man or woman would be like. Experts endlessly debate whether future generations should be tall or short, smarter or average, but come to no conclusion. Geneticists aren't even sure which traits are "bad" enough to warrant sterilization, abortion or the risk of gene surgery.

The poet Byron had a clubfoot. Dostoevski was an epileptic. The electrical wizard Steinmetz was grotesquely deformed. Abraham Lincoln suffered from a debilitating hereditary disease called Marfan's syndrome. Had they been born in a future era, would they be victims of a government screening program?

And how much will racial and political considerations figure in the calculation of perfection? The Nazis insisted that Slavs and Jews were furthest from the Ideal Man. A Russian or Israeli eugenicist would surely think otherwise.

At the core of every eugenics proposal lies an unshakable belief that man can be made perfect. This "philosophy of evolution" is built on the assumption that the human race began as crude savages and gradually developed larger brains, inventing culture, religion and technology along the way.

Progress is god in this world view, and devotion to it, reaching its peak just as Darwin came on the scene, is what fuels the idea of eugenics. It is more than coincidence that the founder of the movement was Charles Darwin's cousin. If we came this far toward perfection, the reasoning goes, why shouldn't we use every technique available to continue toward the supermen we could be?

Genetics is merely the latest in a long string of attempts to perfect mankind. Promising salvation through education, social reform, redistribution of wealth and psychoanalysis, among others, these movements have strewn human wreckage across the pages of history.

In spite of the painful lessons of the past, the cry can be heard today for population control, followed by quality control, followed by the superman. Once the idea is accepted that man should be improved—and a large number of today's scientists do accept it—a full-scale eugenics program is the logical consequence. It may not begin in democratic countries at first, but it could ultimately reach our shores if those who oppose it do not make their voices heard.

The conclusion of Nicholas Wade, senior staff writer for the respected journal, *Science*, illustrates how acceptable the new eugenics is among those on the frontiers of biology:

> Racist beliefs and Nazi attempts to create a master race have given eugenics an odious reputation. Yet *if the historical precedents are laid aside*, the concept may appear to have certain merits. . . .
>
> The concept of progress is a deeply rooted value of industrialized societies: when a means is found to design indubitable improvements in the intellectual or emotional architecture of the human mind, would it be true to that value to ignore the opportunity of improving a scarcely perfect species?[18]

Have the failures and outrages of a century of eugenics been suffered for nothing? Will the god of progress drive us further along the road to destruction? Even while we wait for an answer, another new power is stirring which may bring us to the precipice sooner than we realize.

The Coming of the Clones

Having suppressed all but the smallest twinges of guilty fear, Susan strode purposefully through the gleaming, chrome-bordered glass doors and approached the receptionist. The interior of the Reprotech Clinic was pleasant and warmly lit, friendly enough to put the most nervous clients at ease. Like the commercial abortion centers that flourished back in the 1980s and 90s—in by nine, out by five—the clinic went to great lengths to ease the nagging moral qualms that so many carried over from an earlier era.

Upon learning the reason for Susan's visit, the smiling young receptionist ushered her into a comfortable counseling room. "The doctor will be here in a moment," she chirped brightly, closing the door with a barely audible click as she swished out of the room.

A minute or two later, the door reopened and a short, burly man dressed in a doctor's starched white coat entered, immediately taking the measure of the plain-looking blonde before him. "I'm Dr. Hess," he said ponderously, attempting a friendly but distant smile as he sat down behind the chrome desk. "I understand you're interested in having a baby."

"That's right," Susan replied self-consciously, her voice cracking at first. She smoothed nonexistent wrinkles from her bright blue pantsuit and carefully chose her words. "A friend told me about the clinic and suggested I talk to you about—well, cloning a baby for me."

Dr. Hess gave a slight involuntary jerk at the word "clone." Once a renowned researcher in genetics, he had lost his reputation over illegal experiments involving that very word. A man before his time, he always thought of himself.

"We prefer to call it reproductive duplication," he declared with a condescending smile. "Although it has been legal with humans for five years now, some words still carry . . . ah, unfortunate meanings to the general public."

"Oh, I understand," stammered Susan, unhappy at having begun the interview so poorly. "Well, I live in the city with another woman and . . . you know, we're . . . well, I've always wanted a baby and it's impossible without. . . ."

Dr. Hess held up a hand. "No need to explain," he reassured. "I understand perfectly. We have performed duplication procedures on many people just like you. It's much easier since you're a woman, you know, because you can carry the baby yourself. A man, on the other hand, must have the extra expense of a mechanical womb and such.

"Now," he continued, leaning forward and assuming his most professional manner, "are you familiar with the process of reproductive duplication?"

"Well, I know a little from what I've read. I understand you can take a cell from anywhere in my body and grow my exact twin from it."

"Something like that," the doctor muttered with another quick flash of the superior smile, "although the process is a little more complex.

"Basically, it involves scraping some cells from the inside of your mouth and growing them in a chemical solution until we have plenty of cell nuclei—the part that contains all the DNA that makes you 'you'—to transfer into an egg cell.

"Then, we'll do a simple surgical procedure to remove some eggs from your ovaries and insert the nucleus of one of the body cells into one of the egg cells. After that, we just implant the fertilized egg into your womb, and nine months later you'll have a healthy baby girl."

"Will she look just like me?" Susan asked timidly, worried because the procedure sounded so simple yet so incredible.

"Like a Xerox copy," Dr. Hess winked in a feeble attempt at humor. "It's safe, sure . . . and we take credit cards. Are you still interested?"

"Oh, yes," Susan responded quickly, wondering at the same time what her strait-laced mother would say when she heard about it. The whole thing appeared somehow unholy, like a Frankenstein story. Still, Dr. Hess seemed to consider it routine, and she did so want a baby. . . .

"Good. I'll make an appointment for you," announced the doctor, abruptly breaking into her meditations.

Standing shakily to her feet and grasping the doctor's outstretched hand, Susan managed a thin smile and left the office.

"Have I done the right thing?" she pondered as she

stepped onto the crowded sidewalk. "Oh, well, *everybody's* having them these days," she concluded firmly and hurried to catch a cab.

A Thousand Clones

Whether or not cloning becomes the routine procedure Susan found, the genetic duplication of humans appears to be on its way. Few subjects capture the popular imagination as quickly or consistently as the growing of human doubles. The prospect of a thousand Einsteins exploring the mysteries of advanced physics, enough Marilyn Monroes to satisfy the desires of every woman chaser and a room full of Beethovens composing masterful new symphonies seems to attract the fancy of the public more than any other aspect of genetic engineering.

Boys From Brazil, the story of an attempt to unleash a group of Hitler clones on an unsuspecting world, brought the subject to movie theaters, and its success spawned many film and television imitators.

Soon after the untimely death of superstar Elvis Presley, the celebrity tabloid *Modern People* claimed to have discovered solid evidence that a genetic duplicate of the singer was born eight months before his death.[1]

Allegedly performed at a secret laboratory on Guam, the operation was reported to have cost ten million dollars. "Elvis knew he was going to die," insisted one of the unnamed sources, "and wanted to leave a memory of himself on earth." Judging by the popular success of not-too-convincing Elvis imitators abounding today, such a clone could expect guaranteed stardom—beginning around 1994!

The improbable story of the Elvis clone appeared one month after the publication of David Rorvik's controversial bestseller, *In His Image*, and was perhaps inspired by it. Rorvik, a freelance science writer who had published several articles on the subject of genetics, tells of being approached by a sixty-seven-year-old bachelor millionaire who wanted a clone child to carry on the family name. A long search for a doctor willing to attempt the operation turned up a man he appropriately calls "Darwin."

Although skittish and temperamental, Darwin eventually establishes a laboratory in a remote tropical country "beyond Hawaii." There, with the assistance of other anonymous researchers, the doctor conducts his cloning experiments.

Following a disappointing series of failures, a beautiful young native girl, code-named "Sparrow," is impregnated with the millionaire's clone, allegedly giving birth to a healthy baby in 1976 as a "fitting contribution to the Bicentennial." Possibly to remove the racist overtones of the book—local non-white girls being used for the experiments, paid handsomely by the paternalistic American researchers isolated in their jungle enclave—Rorvik closes his book with hints that Sparrow and her millionaire are in love, perhaps to be married.

Since the procedure was performed·in secret, with all the participants remaining anonymous, questions immediately came up about the truth of Rorvik's account. Conceivably to avoid lawsuits (unsuccessfully, it turned out), the publisher added a disclaimer to the book: "The author assures us it is true. We do not know."

Further investigation revealed that maybe they should have known. Rorvik had originally approached Dr. Landrum Shettles (the doctor in the Doris Del Zio test-tube baby case) in 1976 with the suggestion that he help with the cloning of a multimillionaire. Although Shettles showed interest and began preliminary work on the project, he heard no more from the author.[2]

The following year, Rorvik wrote prominent Oxford University geneticist J. Derek Bromhall, requesting information on cloning to be used in a book. He received in reply a nine-page abstract of Bromhall's doctoral thesis on the subject. To the scientist's surprise, the information he supplied formed the basis of *In His Image*.

More incredible, the alleged birth of the clone baby occurred five months *before* Rorvik contacted Bromhall. If the world's first human clone had been created in Rorvik's presence, why would he request an outdated scientific paper on the subject? Bromhall, asking the same question, filed a seven-million-dollar suit against the author and his publisher.[3] To date, Rorvik has supplied no shred of evidence that the cloning actually took place, and no researcher in the field believes it has been done—except Dr. Shettles.

Still, Rorvik received the publicity he craved, and interest in the subject continues to run high. No doubt fed by an uneasy feeling that duplication of humans is wrong, most of the cloning fantasies involve a self-centered manipulator such as Rorvik's millionaire or an evil genius like the Nazi who supervises the multiple cloning of Hitler in *Boys From Brazil*.

The Coming of the Clones

Why Clone?

Although widespread human cloning remains perhaps a generation away, the subject is important because it dramatizes the possibilities and problems brought by the new biology. Writes Dr. Willard Gaylin, president of the Institute of Society, Ethics and the Life Sciences, "Many biologists, ethicists and social scientists see cloning not as a pressing problem but a metaphoric device serving to focus attention on identical problems that arise from less dramatic forms of genetic engineering that might slip [unnoticed] into public use."[4]

Whether cloning is a future fantasy or a pressing problem, why would people want to create human duplicates, anyway? From the visionaries and the pragmatists, a variety of reasons come forth.

Those interested in parents' desires to have children look to cloning as a means by which any infertile couple can reproduce. Since a clone can be generated from either the husband or wife, the biological offspring of at least one parent could be born to any family—perhaps alternating parents with each child so both could feel fulfilled. Aunt Martha's delighted exclamation, "Why, he looks just like his father!" would take on new meaning if a boy were a carbon copy of his dad.

In cases where a couple is fertile but carry a serious genetic disease such as diabetes, they might have a clone child rather than run the risk of transmitting their illness. The baby would also be a carrier, but would be at least as healthy as the parent from whom he was taken. And a side benefit of any cloning process is the automatic control of the child's sex: from the father comes a boy, from the mother a girl.

To those more interested in re-engineering society, the coming of the clones offers a golden opportunity to experiment and devise a new social order. Stanford's Dr. Joshua Lederberg hopes it can be used to "clear up many uncertainties about the interplay of heredity and environment; and students of human nature will not want to waste such opportunities."[5]

Scientists see the establishment of a clone colony, made up of identical twin groupings, as a perfect model for studying the development of social harmony. The potential for misunderstanding would be greatly reduced—some even predict a form of ESP among clones—and cooperation enhanced in this colony of mirror-image citizens.

Building on this thought, a few foresee the development of specialized clones who might be designed for particular occupations. A race of quick-thinking, rugged soldiers could defend us, while another strain of blue-collar clones, completely happy with the monotony of an assembly line, might manufacture our automobiles, washing machines and television sets at reduced cost.

Frozen clone embryos could be rocketed into space, with only a handful of their fully grown brothers on board to pilot the ship. Sent to colonize distant planets, they would ship back needed raw materials to earth. Since all the clones are duplicates, organ transplants would be simple—free of tissue rejection—and each colonist could live for centuries.

Back on earth, those interested in fulfilling for a price every individual's fantasy might supply clones of favorite celebrities or great people from the past.

Judging by the almost religious reverence attached to pictures and memorabilia of Elvis Presley and other stars, it is not difficult to imagine a woman paying a fortune to bear the clone child of her idol. Armed guards around the graves of notables could become commonplace to prevent the theft of tiny scrapings of a celebrity's body in order to feed a black market in cloning.

Along the same lines, it would be possible for any of us with enough money to become biologically immortal through cloning. A clone could be kept in storage as a "spare parts bank" for each individual. Prevented from developing a brain, it would not be considered a person, and surgeons could draw on its organs to repair our failing bodies indefinitely.

Famed geneticist J.B.S. Haldane long ago proposed that people especially important to society retire at age fifty-five and spend the rest of their lives raising and teaching their carbon copy offspring.[6]

Might not a world populated by generations of clones be sufficient cause for God to sweep the earth with another universal destruction? How far will He allow man to go in manipulating Creation? And more immediately, how did we come as far as we already have?

Cornell Carrot and Oxford Frog

Like most advances in the scientific understanding, cloning approached in tiny steps, its ultimate potential far from the researchers' minds. The first clone experiments occurred around the turn of the century, using simple animals such as sea urchins and worms. In 1939, readers of *Life* were startled when its cover featured a cloned rabbit.

But this was not true cloning. Instead, it was induced "parthenogenesis," or virgin birth, a process related to cloning but much cruder. By subjecting unfertilized eggs to stress—dry ice in the case of the sea urchin, heat shock with the rabbit—scientists stimulated them to reproduce asexually (having only one parent). In the 1950s, for example, researchers reported nearly four hundred different ways of causing sea urchins to reproduce without the help of a father.[7]

Stories persist of human parthenogenesis as well, in which a sudden shock to a woman causes her spontaneously to become pregnant. London University College geneticist Dr. Helen Spurway estimates that one in every 1.6 million pregnancies might occur in this way, and a few such cases have been documented (although not to everyone's satisfaction).

The most famous incident involved twenty-year-old Emma Marie Jones, who collapsed from shock in the streets of Hanover during a 1944 Allied bombing raid. Nine months later, she gave birth to a daughter, Monica, whom English investigators judged to be her exact genetic duplicate. Extensive tests failed to disprove her claim of not having had sexual relations, although a test skin transplant from mother to daughter surprisingly failed to take.[8]

For those who might see theological implications in such an alleged oddity of nature, keep in mind that parthenogenesis can produce only *female* offspring. No male (Y) chromosome is involved in the child's conception.

A true clone, in the current use of the word, requires

more than stimulation of an egg cell to be conceived. Modern biology is attempting to reproduce plants and animals by causing a cell from anywhere in the body, rather than the normal sex cell, to grow into a full-scale genetic copy of the parent.

The process is based on the discovery that our complete genetic code is contained in every single cell of our bodies. Through some means as yet unknown, only a portion of the DNA blueprint—the "armness" in an arm cell, or the "braininess" of a brain cell—is switched on. This is called *differentiation.* If a cell can be chemically persuaded to "de-differentiate" and turn on all the banks of its genetic computer, it could grow into a carbon copy of the complete original—a clone.

This incredible feat was first accomplished with plants in the early 1960s by Dr. Frederick C. Steward at Cornell University. After working for several years, Dr. Steward finally hit upon the right combination of fluids which would stimulate a mushy broth of orange-red carrot cells to "de-differentiate" and reproduce themselves into complete carrots. The results were dramatic as the cells rapidly turned green, multiplying eighty-fold in weight in the first twenty days.

Based on Steward's successful efforts, commercial agriculture now routinely clones chrysanthemums and other plants for sale to the public. The next serving of strawberry shortcake you eat could well contain cloned strawberries, which make up a significant proportion of the crop.

Says one leading scientist, "The leap from single cell to cloned carrot is greater than the leap from cloned carrot to cloned man." But before reaching man, someone first had to clone an animal, a much more difficult procedure than working with plants.

This was done by British biologist Dr. John B. Gurdon, who removed the nucleus of frog body cells—from the intestinal wall, tail or anywhere else in the frog's body—and inserted them into unfertilized egg cells. Although the operations were incredibly difficult, some of the eggs matured and grew into full-sized frogs.

Preliminary successes also have been reported with mice, but no true cloned mammal has been produced to date. Gurdon's frog experiments are as far as we have come, partly because mammal eggs are only one-tenth to one-twentieth the size of frog eggs and partly because mammal embryos require a womb in which to develop. With continued advances in microsurgery and progress on an artificial womb, however, few scientists doubt that mammal clones will also become a reality.

The Coming of the Clone Child

What about human clones? Is this merely the dream of science-fiction writers, or are we on the threshold of a major breakthrough? While the cloning of a human being won't be easy, there is no reason why it could not be accomplished when the procedure is perfected with mammals.

"Once you get the chromosomes from a body cell in [the egg]," observes Dr. Kurt Hirschhorn, chief of medical genetics at Mount Sinai School of Medicine in New York, "there's no reason whatever why [it] won't grow just like an ordinary fertilized egg cell."[9]

Of course, technical difficulties and other problems

must be overcome. So far, only body cells from very young animals and plants have been used in successful cloning experiments. It may be that with age, the body cells become so firmly set in their particular role—part of an eye or an intestinal cell—that they could not successfully develop into a complete organism even if they were "switched on."

And always there is the question of failures, especially significant with human clones. In Dr. Gurdon's frog experiments, three-quarters of the eggs failed to reproduce. Of the ones that did divide, many grew into tadpoles with damaged eyes or twisted tails and intestines. Because the potential for mistakes is so great—even the tiniest excess in suction or pressure can damage the cell when it is being injected—only eleven clones grew from 707 attempts in Gurdon's experiments.

How Would the Clone Feel?

Although there seems to be no reason in principle why human cloning could not be developed, there are strong moral reasons why, in the words of embryologist Robert T. Francoeur, "it shouldn't be done . . . even once."[10] Because the laboratory procedure is so chancy, the vast majority of "bench embryos"—a term commonly used by researchers for human beings in the early stages of development—would have to be "jettisoned," or flushed down the drain.

This casual abortion of human embryos, whether by destruction before implantation in a womb, spontaneous abortion or abortion after amniocentesis, is an integral part of the cloning process. In the minds of some scientists, today's approval of human abortion on demand provides the solid rationale for cloning experiments tomorrow. Asks Dr. Douglas Bevis, ridiculed in 1974 for his unsupported claim to have produced the first test-tube baby: "How can any society that accepts termination of pregnancy quibble about giving life to a fetus? We are not creating life. If we can kill the fetus—and this seems to be expected and accepted—why can we not 'put it together'?"[11]

And what about the clone himself? While society might consider him an alien species of some kind, a human clone would still be a person, with feelings, desires and an understanding of his uniqueness. As one writer warns, "Clonal reproduction introduces something totally new into the world—the mind of a child who *knows* it is a biological replica of its parent, a child who knows it is largely preordained, a freak who can see its biological future mirrored in another person."[12]

A clone child would probably be, more than any other youngster, the product of the parent's fantasies, expected to fulfill the dreams he could never realize. As the parent tries to develop a true duplicate of himself, the child's own potential will be stunted and his outlook warped as he is forced into a mold he neither fits nor wants.

Theologian Paul Ramsey points out that growing up as a natural twin is difficult enough without the additional burden cloning would bring:

One's struggles for selfhood and identity must be against the very human being [his twin] for whom no doubt there is also the greatest sympathy. Who then would want to be the son or daughter of his twin? To mix the parental and the twin relationships might well be psychologically disastrous for the young.[13]

What would be the future of the child replica who fails to meet his parent's high expectations? After all, he is only a biological copy of the parent, not a psychological or intellectual reflection. What of the duplicate Einstein who, far from pioneering in mathematics, turns to alcohol as an escape from the pressures on him and winds up in some skid row alley? Doesn't society produce enough misfits already without designing new ones?

The resentment clones feel against their parents and the society that allowed them to be created might turn them into the disillusioned terrorists of the twenty-first century. Roaming like the bands in Stanley Kubrick's *A Clockwork Orange*, these clone gangs could add a new dimension to urban fear.

A Right to Be Unique

What is a human clone, anyway, but a person forced against his will to duplicate a life already lived? Doesn't cloning imply ownership by the parents which denies the child's freedom to develop his or her own nature? Parents are temporary custodians of an original individual, uniquely created by God.[14] What right do they have to smother the singular divine acts by molding a creation in *their* image?

The divinity and worth granted every human being is fearfully jeopardized by the threat of genetic duplication. Born into a special relationship to our Creator, each of us is free to develop or reject this spiritual bond as we spend our brief time in this life. A human clone would be deprived of uniqueness, uneasily approaching God as if a fragment of a split personality.

Being human in every respect, a clone certainly would possess an individual soul, just as he would have an eternal destiny separate from all others. Mere genetic duplication does not condemn a person to an unalterable future, any more than natural twins (also genetic duplicates) have to follow the same course in life. Still, a clone would be likely to feel a spiritual shadow lying over him, perhaps in his own mind darkening his right to communion with God.

Psychologically and spiritually, then, the human clone is at an incredible disadvantage. Even beyond theological objections, this crippling places any such attempt outside the limits of morality.

It is questionable whether society needs or can cope with human cloning, anyway. Says Dr. Theodosius Dobzhansky of Rockefeller University, "It can show no lack of respect for the greatness of men like Darwin, Galileo and Beethoven, to name a few, to say that a world with many millions of Darwins, Galileos or Beethovens may not be the best possible world."[15]

With today's interest in expanding human rights and individual freedoms, the last thing we need is citizens brought into the world with their lives predetermined. And with sexual identities blurring and homosexuality flaunting itself as the wave of the future, a process allowing a homosexual to raise a clone family would bring more ruined lives. The fabric of modern society is already torn enough without introducing such massive new rents into its tattered weave.

Cloning Around the Corner

Can cloning be stopped? Duplication of mammals is prohibited for U.S. scientists receiving Federal funds, but how effective is this ban in the long run?

Since the successes of Steptoe and Edwards with test-tube babies, the manipulation of human embryos has taken a great leap forward, and implanting an egg into the womb is now a proven procedure. The main obstacle to human cloning is fertilizing the egg with a body cell, a problem that might be solved with new techniques under study.

The most promising of these, fusion, is receiving a great deal of research attention because it now offers one of the best avenues for understanding the cause of cancer. In addition, it is the object of intense study by those attempting to unravel the biochemistry of diseases such as cystic fibrosis and multiple sclerosis. Thus, without even thinking of attempting a human clone, researchers are steadily perfecting the techniques.

As the cloning of mammals approaches, the demand for veterinarians to use the procedure—perhaps the secret duplication of a champion race horse such as Secretariat, where millions of dollars are at stake, or the mass cloning of cattle and chickens to increase efficiency and reduce prices—might prove unstoppable. Even Oxford's Dr. Bromhall, a conservative scientist who maintains that no sane person would ever attempt human cloning, believes that the duplication of domestic animals is "desirable and will ultimately prove feasible."[16]

As animal cloning reaches new triumphs and a demand develops for human cloning by the infertile or the growing homosexual movement, what is to stop the attempt? One man's mad scientist is another's dedicated researcher.

Dr. Shettles, embroiled in both the Del Zio test-tube baby and Rorvik cloning incidents, recently became involved in another controversy when he claimed to have grown a human clone to the implantation stage in a Randolph, Vermont, hospital. Announced in the respected *American Journal of Obstetrics and Gynecology,* the experiments raised some eyebrows because of sloppy research work. Yet if the report is true, it means that human cloning is almost here.[17]

Cloning experiments need not be expensive, experts say, with the smallest of countries possessing the resources for eventual success. Somewhere in the world, a dictator, millionaire or madman will commission willing scientists to bring to birth the first human clone. If the procedure is successful, a rush to duplicate the effort will no doubt follow.

How far away is this day of destiny? If we were to believe the predictions of some scientists in the early 1970s, cloning should be occurring right now. More recent forecasts range from "five or ten years" to "not in the foreseeable future." But the majority believe it will take place before the end of the century.

Commenting on the likelihood of cloning, DNA co-discoverer James Watson concludes: "No doubt the person whose experimental skill will eventually bring forth a clonal baby will be given wide notoriety. But the child who grows up knowing that the world wants another Picasso may view his creator in a different light."[18]

We are well along in development of many shocking

abilities to reshape the meaning of humanness. Some of these powers are good, some are evil. Who will sort out one from the other? Will government control the breathtaking advances of our brave new world? Industry? Science itself? Or do these new skills contain the seeds of our doom?

10

Closing Pandora's Box

It is the twenty-first century. World leaders have come to realize the need for drastic moves in order to save our wounded civilization.

The steady decline of energy resources slowly saps the world's lifeblood, while industrial innovation stagnates under the terrible twins of rampant inflation and oppressive taxes.

As the often-patched social fabric rapidly unravels, crime rages through the world's great cities. Children scorn their parents, and a drug-induced haze is all that carries many citizens through each new day.

In desperation, leaders are turning to genetics as their final hope. At first, the Western democracies were reluctant to begin genetic re-engineering of their populations. But disturbing reports from the Soviet Union finally spurred them into action.

The Russians had launched a nationwide program to transform their whole society within a generation, raising intelligence and stamina levels through mass injections of genetic material. Babies have been modified in the womb according to a complex master plan that predetermined the role each new citizen would play.

Intelligence sources reveal that the Soviets also have been breeding super-soldiers, genetically designed for strength, quick thinking and coolness under fire. Cloned in the thousands to populate an unbeatable Red Army, the vanguard of this force are just entering their twenties, a fearful new threat to the balance of power.

In the West, the media rivets public attention on the growing "gene gap" with the Russians, forcing leaders to activate their own genetic programs to keep from falling further behind. The space race was nothing in comparison with the new gene race.

Among the democracies, frantic re-engineering of the population is beginning in order to meet the Russian challenge. Giant computers link the Western world under the control of one all-powerful leader. Individual and national freedoms take a back seat as leaders realize that their societies are in a struggle for their very existence.

Citizens who resist genetic redesign on moral grounds are branded as unpatriotic, their children torn from them to be raised by the state. Plans are being made to tattoo each person's genetic code on his forehead and hand with invisible electronic ink, to speed mobilization against the coming threat.

And so the world moves anxiously through the century, turning the promised benefits of biotechnology into fear and slavery. Many wonder how long mankind will survive this mass manipulation of its humanity. Will God intervene to save us from ourselves? Fewer and fewer cling to this ancient hope as the slide into darkness continues, driven by an apocalyptic sense of impending doom.

The Russians Are Coming

This futureworld scene may forever remain a distant nightmare, but it rests solidly on today's developments. The revolution in biology is international in scope and on the verge of becoming fiercely competitive. With public debate on the subject largely limited to the United States, labs in Europe and Russia move quickly to explore the secrets of life. Can one nation resist developments believed to be wrong when "They" are pursuing them?

Past experience with germ and chemical warfare and the development of nuclear weapons supplies the sad answer. As the genetic age bursts upon us, the pressure to develop new technologies steadily builds. In the United States today, a hundred universities and a dozen private companies race ahead with DNA research.

According to our best evidence, reported a Congressional inquiry, there are virtually no limitations on genetic research in the Soviet Union, while American efforts have been sharply restricted. Scoffs Vladimir Engelhardt, director of the Soviet Institute of Biology, "Soviet biologists . . . believe the ethical side of the problem has probably been exaggerated. . . . When genetic engineering becomes possible, society will be mature enough to overcome the possible dangers."[1]

Another member of the Soviet Academy of Sciences, Dr. A. Neyfakh, predicts that a gene race will soon be launched by the West to overcome a supposed "brain drain" among the capitalist nations. He expects us to engineer new mental powers among our citizens and urges his fellow Russians to jump the gun and be first with the new technology.[2]

Jockeying for the leadership position in this lucrative new field, major American corporations are also applying pressure to ease restrictions and allow them to plunge ahead into genetic technology. The right to hold exclusive patents on new life forms makes entry into this realm irresistible for the companies, with licensing fees alone expected to bring in multiplied millions of dollars.

Adding together the pressure of the marketplace for new products and the threat of international competition, it seems that moral qualms about new developments are headed for second place. The most profound changes in our future life may be the result of industrial and political pressure, rather than any carefully weighed, cautious advance into a sphere we barely understand.

Bottling the Genie

Can anyone control the genetic juggernaut? The attitude of most scientists is that they will police themselves, thank you, and the public should stay out of it. Leaving aside the question of how scientists, most of whom work directly for industry, would perform this policing, what is the likelihood of effective controls from this quarter?

An admirable first step was taken when an international conference met in 1975 at Asilomar in California to discuss the hazards of the new genetics. Unprecedented in the history of science, it was held to consider limiting for the first time the freedom of scientists to investigate a new field.

As a result of the conference, safety guidelines were set up to control some aspects of biotechnology. Much remains unregulated, however, and many scientists flaunt the existing guidelines because they just don't believe in them.

As an example, science writer Janet Hopson spent three months observing operations in a leading university genetics lab and watched in fascinated horror as workers smoked and ate while working on dangerous organisms, often sucking them into glass tubes with their mouths. She asked one PhD. candidate how the safety guidelines applied to the organism he was working on; he hadn't the faintest idea. When the lab was temporarily shut down because of these conditions, the closure was treated jokingly by the workers, few of whom knew or cared much about safety guidelines.[3]

Such reports from the real world of science, in contrast to the exaggerated look of concern many scientists wear when testifying at public hearings, confirm the prediction of DNA co-discoverer James Watson: "There should be no illusion that regulation is possible."[4] And if safety is of little concern to the scientists, how can moral qualms be expected to stop their work?

Is Freedom Absolute?

Science sets its own pace and direction, we are told, and it will be followed like the Pied Piper wherever it leads. Cut off tax funds and the researchers will go to private industry. Regulate industry and they'll move to other countries. Scientists consider their freedom to be an absolute (while calling everything else "relative," it might be noted), a measure of civilization itself.

We tend to forget that scientists are human beings first, with all the failings and desire for glory that characterize our race. No one doubts the benefits science has brought us, but we have a right to question those who set it up as a kind of deity, its siren call to be pursued wherever it leads.

Scientists have become wary of any attempt at public scrutiny, fearful of an anti-science movement that gathers power each year. When on the brink of control over life itself, it is natural for people to raise more questions than ever before. But the scientists' response has been to fight every inch of this intrusion into their home ground.

Reports author June Goodfield, "Whenever I have talked to scientists about [genetic engineering], I have found that their attitudes seem to be governed partly by whether they think their interlocutors are for or against science." If you appear approving, she explains, they will excitedly tell you how promising the field is. But should your attitude seem antagonistic, the response will be to "emphasize that these kinds of applications are so far in the future . . . and so highly complex that they are really irrelevant to [today's] issues."[5]

National Cancer Institute biochemist Dr. Maxine Singer goes so far as to insist that genetics research must continue because "nothing less than science itself is at stake."[6] Perhaps it makes more sense to question that very research because nothing less than mankind's future is at stake.

While important, science is just one among a number of attributes that make up our world. Religion, ethics, art and other values deserve equal consideration with the imperatives of science.

Why should we leave the future direction of life on this planet to science's self-policing efforts? On these questions, the judgment of scientists is no better—and because of the blinders of narrow specialization, is probably worse—than that of the rest of us. After all, *we* are the ones who must pay dearly for any miscalculations.

Government Green Light

If science cannot be relied upon to control the new biology, what about government? Will our elected representatives weigh the moral and social costs of biotechnology before allowing it to proceed?

Hopeful signs do exist that some levels of government are taking an interest in these new developments. The city council of Cambridge, Massachusetts, held widely publicized hearings in 1977 before allowing dangerous genetics research to continue at Harvard. House and Senate committees have investigated the subject extensively in the past few years. An HEW Ethics Advisory Committee to study the question of test-tube babies began work in 1978 and held hearings in cities across America to gauge public and scientific opinion.

But the results so far are disappointing. The government's general response has been to give biotechnology a green light and pour out financial blessings upon it.

In the United States, only the Federal government has any real power to control or channel this research. On the whole, its attitude is that since the research will continue anyway, only minimal regulations should be fashioned. Having no real policy on biotechnology, the government will continue to make superficial reviews of new discoveries, coupled with steady increases in research funds.

Significantly, the question of morality comes up only rarely in government discussions. In more than two thousand pages of Congressional hearings on regulating genetics research, it appeared primarily in the form of witnesses and Congressmen reminding themselves not to forget ethics—and then promptly forgetting it.

Senator Edward Kennedy, the driving force behind the Congressional review of policy in this field, said, "From a Constitutional point of view, the frontiers of law for the next twenty to twenty-five years will be in these areas of bio-ethics."[7] Still, his own committee virtually ignored the moral issues.

Congressional investigators have concluded that benefits outweigh the risks and that tax money should be spent to advance it "in as positive a fashion as possible." The burden of proof about dangers, they declared, *lies with those who oppose the research*, not with the scientists conducting the experiments.[8]

Closing its report, a House subcommittee stated, in a chilling parallel to the scene that opened this chapter, "Any significant U.S. lag in genetic technology, for whatever reasons it may occur, could result in diminished power and prestige for this nation on the international scene."[9] They recommended that the already-minimal safety guidelines be relaxed to spur further research, a suggestion put into effect in 1979.

The conclusions of HEW's Ethics Advisory Board regarding test-tube babies were also small comfort. Following months of hearings, the board managed to sidestep nearly all the difficult moral questions and agreed that test-tube research should continue with few limitations.

Under the board's recommendations, experiments with human embryos are permitted once again (they were banned in the United States in 1975), whether the purpose is to aid infertile couples or just general research. The only requirement is that embryos used for experimentation not be kept alive more than fourteen days.[10]

In fatalistic tones, the committee draft declared that "the development of this technology in the private sector appears to be inevitable," and the best the government can hope for is to guide the research somewhat. The thought of stopping it altogether was hardly considered.[11]

It's Up to Us

If the government should become serious about controlling the new biology, its present powers extend only to tax-supported laboratories, anyway. Baby Louise's birth came through privately supported research, and as royalties from new patented life forms begin rolling in to companies and universities around the world, government financing will fade as a factor in control.

With the lid to Pandora's box now standing wide open, and science, government and industry unwilling or unable to close it again, the public is left to its own resources if effective controls are to be placed on the new technologies. Concerned public opinion—with the church in the lead—will have to confront those charged with regulating this new field of science if moral and safety objections are to be given the consideration they deserve.

Notes Dr. Ruth Hubbard of Harvard's Biology Laboratory, "[Scientists] imply that there is an enormous hurry; that something terrible will happen if these experiments are not done immediately. Something terrible will happen only if a dangerous experiment *is* done, not if an innocuous one is delayed for awhile."[12]

In general, ethicists, theologians, philosophers and politicians have been reluctant to question the advance of science. But the stakes are so high and our future so cloudy that we who respect God's Creation cannot sit idly by any longer. Greater principles are involved than freedom of scientific inquiry, and we will not discover them in corporate board rooms or Congressional hearings. It is now a question of whose plan for humanity we will follow: God's or man's.

11

In Whose Image?

The California sun beamed brightly through the nave of the suburban church where Roger Meyers was at last marrying the girl of his dreams, Virginia. Radiant in a white satin gown with a three-foot train, the smiling bride felt twinges of nervousness at seeing such a large crowd, but Roger's presence next to her eased the fears. After all, he was the rock of determination who had overcome all obstacles to bring this special day about, working tirelessly as a restaurant busboy to earn the money they needed to begin married life.

A typical June wedding? Yes, except for one major difference. Both Roger and Virginia are mentally retarded, and they spent long years learning to manage such simple tasks of married life as shopping, balancing a checkbook or reading bus schedules. During that time of breaking free from the institutional mold, only the two of them believed the wedding would ever take place.

A decade ago, it wouldn't have. According to experts, the pair would have been warehoused for life in large state facilities, sterilized without their consent, fre-

LET US MAKE MAN

quently drugged for easier care and allowed little or no contact with the outside world. Now, with a shift in public attitudes, a growing number of the retarded and handicapped are overcoming years of neglect and doing things once unthinkable.[1]

Sadly, this ray of sunshine through dark clouds of intolerance may be a last flash of brightness in the twilight. With the arrival of genetic screening and the emphasis on producing perfect babies at any price, society may no longer tolerate the birth or continued life of people such as Roger and Virginia Meyers. In its quest for the New Man, our culture could well find these "defectives" too great a hindrance to tolerate.

Into a New World

Has science now brought us to the day of which our ancestors dreamed when they built the Tower of Babel, when "nothing that they propose to do will now be impossible for them"?[2]

We have arrived at a crossroads of history, from which most of the signposts have been taken down. Once guided firmly by biblical principles, our civilization now rushes headlong toward freedom without limits. Faced with the greatest moral decisions in history—the ability to shape life itself—people are bewildered and lacking in direction.

Many of the techniques described in this book are in the early experimental stages and may never be developed. But the potential is there. Observes noted science writer Albert Rosenfeld, "Feats that seem purely speculative have a way of suddenly being upon us sooner than anyone had expected."[3] We have looked at numerous examples of yesterday's wild dreams becoming today's headlines.

While scientific knowledge doubles every few years, applications of moral principles sometimes take decades and centuries to develop. Yet we no longer have the luxury of waiting so long. Unless a moral stand is taken *before* the discoveries of science are put into general use—an interval sometimes measured in months—we are left chasing a train that has long since left the station.

The experts agree that the genetics revolution will make the human future clearly different from its past. The question is, which developments should be accepted—both by society as a whole and on a personal level—and which refused?

As Australian biologist D. Gareth Jones points out, "The human race is heading at alarming speed into a totally unknown and unexperienced realm where man himself becomes the controller and potential manipulator of his own body and brain. . . . This new biology will raise, and has even started to raise, questions with far-reaching implications, chief amongst which must be, 'What is man?' "[4]

A Certain Likeness

What is man? Where did we come from and where are we going? Our answers had better be clear, for they will shape our response to the challenges of biotechnology and determine the kind of world our children grow up in.

The biblical reply rings through the centuries: "God created man in his own image, in the image of God he created him; male and female he created them."[5] Not only is this the foundation of Christianity, but it is the sole unwavering source of human dignity and value.

While the resemblance between God and man is not a photographic one, we share a likeness in essence, especially in our personhood. We are warned in the Bible not to shed another's blood or curse our fellow man *specifically* because each of us is made in the image of God.[6]

Our created likeness to God also means that each of us is a unified whole, not merely an animal with reasoning powers or a religious sense tacked on. Even the traditional distinction among body, soul and spirit made by theologians is not an actual separation but a division for purposes of discussion and understanding.

We do not receive a soul or spirit at some point in prenatal development; we *are* soul and spirit as well as body. As the great theologian G.C. Berkouwer explained, "The whole man is made in the image of God, . . . [not just] certain 'higher' qualities." Renouncing God, Berkouwer maintains, involves a rejection of man's personhood as well.[7]

This lofty understanding has formed the core of our civilization, a view that demands deep respect for human life and compassion on those who may be physically or mentally imperfect. Famed biologist-philosopher Jean Rostand, while not a Christian, is one of the great thinkers influenced by this ethic:

> For my part, I believe that there is no life so degraded, deteriorated, debased or impoverished that it does not deserve respect and is not worth defending with zeal and conviction. . . .
>
> Above all, I believe that a terrible precedent would be established if we agreed that a life could

be allowed to end because it was not worth preserving, since the notion of biological unworthiness, even if carefully circumscribed at first, would soon become broader and less precise. After eliminating what was no longer human, the next step would be to eliminate what was not sufficiently human, and finally nothing would be spared except what fitted a certain ideal concept of humanity.[8]

That "terrible precedent" of unworthiness is well on the way to being set. As thinkers such as Rostand continue to chip away at the spiritual foundations of human value, they are shocked and dismayed when others, less influenced by the tradition which calls human life sacred, carry the new biology to its logical conclusion.

Genetic Robots

One leader of the new thought is Harvard biologist Edward O. Wilson. A soft-spoken scientist, he caused an uproar by implying that we are little more than robot-like products of our genes. Basing their understanding of man on evolutionary biology, Wilson and scores of others give a chilling foretaste of where we may be headed.

Whatever the merits of the theory of evolution—a theory that would be much more hotly debated today if it were not such an article of faith in modern science, a trophy of its hard-won victory over the religious establishment—its use as the basis of modern thought leads directly to views such as Wilson's.

Man, he maintains, is just another animal, and questions of morality, religion and behavior should be answered by biology. In general, his conclusions come from studying lower animals and applying their behavior patterns to man. This controversial field is called *sociobiology.*

Since morality evolved as instinct, Wilson asserts, and is determined by genetically programmed centers deep within our brains, we will soon be able to control our future evolution by manipulating these centers. "To chart our destiny means that we must shift from automatic control based on our biological properties to precise steering based on biological knowledge," he insists.[9]

This forms a powerful argument for design of a New Man, and indeed is really an updated version of the old eugenicists' theories. If we are the random product of evolution, scientists wonder, why shouldn't we take matters into our own hands and design the future?

For this, the new genetics provides the key. According to sociobiologist Richard Dawkins of Oxford University, "Humans, animals and other organisms . . . merely represent 'throwaway survival machines' that the selfish genes use to guarantee their own survival. . . . Our genes swarm in huge colonies, safe inside lumbering robots, sealed off from the outside world, communicating with it by tortuous indirect routes. . . . They created us body and mind; and their preservation is the ultimate rationale for our existence."[10]

Could a lower view of humanity be imagined? Certainly, it is considered extreme today, but this outlook is the direct result of replacing the image of God with belief in a blind, random development. Losing the divine pivot point from which all meaning derives, modern man is left floundering in a swampy sea of emptiness.

In Whose Image?

What's It All About?

Ideas themselves are meaningless if we are just programmed by our genes and modified by our upbringing. As C.S. Lewis pointed out, "Whenever you know that what the other man is saying is wholly due to his complexes or to a bit of bone pressing on his brain, you cease to attach any importance to it. But if naturalism [i.e., chance evolution] were true, then all thoughts whatever would be wholly the result of irrational causes. Therefore, all thoughts would be equally worthless. . . . If it is true, then we can know no truths."[11]

Without relationship to God as the Source, we hardly even know what "human" is. One expert lists the will to live and fear of death as the essential qualities of being human, while another sets up a checklist of fifteen requirements. Echoing this thought, ethicist Joseph Fletcher answers the question, "Which would be human, an ape with the intelligence and sensibilities of a man, or a man with the capabilities of an ape?" by siding with the ape![12]

Of course, if man is just an intelligent animal (or a smart set of chromosomes), why not use the less developed of our brothers and sisters for experimentation? It has been done too often already.

While a good deal of the feeling remains from our Christian heritage that man is somehow more than just another animal, this is for most scientists merely a sentiment, lacking any solid foundation. Trying to find words to explain why unborn babies should be treated with more respect than mouse cells, for example, University of California biologist Dr. Clifford Grobstein says lamely, "I guess you might call it a kind of clannishness, . . . a sense that there is something about

147

the human species that is special to us [since] we belong to [it]."[13]

Having discarded the sacredness of man based on relationship to God, modern biology is running head-long in the direction of the genetic robot. The door is wide open for human cloning, manufactured babies, programs to develop a super-race and the rest. These efforts might be halted by economics or sentimental attachments to the old ways, but without a divine rela-tionship at the heart of human worth, secular morality is unlikely to stand in their way.

Human Being or "Tissue"?

The new outlook has been creeping into medical ethics, most visibly with abortion on demand, free from the slightest pangs of conscience. Observes noted surgeon C. Everett Koop, "In talking on rounds to medical students who have never known medicine when abortion was illegal, I find that they have an entirely different concept of the worth of human life—it's cheap. I tell them that when I was in their place the very word 'abortionist' was a loathsome thing; now the abortionist is likely to be the professor of obstetrics in the medical school."[14]

Many Christians consider abortion to be permissible in limited cases, after much soul-searching and prayer. But the whole thrust of today's abortion-on-demand system, now claiming one and a half million fetal lives in the United States each year, is to convince people that an unborn human being is just a piece of foreign tissue—or at the extreme taken by Joseph Fletcher, "a venereal disease."[15]

With such contempt for a baby when unwanted, how

can we expect reverence for the same fetus when it is a wanted child? What is to stop the most dangerous and radical experiments of the new genetics on these "pieces of tissue," especially when they are considered the product of mindless evolution in the first place, easily replaced in nine months if something goes wrong?

Today's attempts to dehumanize unborn children in order to justify their elimination are laying the foundation for a genetic nightmare in the future. We may wake up to find that we have unwittingly accepted every extreme of biotechnology, just as the early eugenicists were shocked when their theories were used to form the heart of Nazi genocide.

On the other hand, valuing unborn children as the image of God on earth also has implications for the new biology. The cloning of human duplicates for any reason is ruled out, as are attempts to design the superman. Genetic engineering to cure disease in the womb is acceptable only if done with extreme caution and reverence for the life at stake. And errors in test-tube conception cannot casually be flushed down the drain as if they were garbage; a moral balance needs to be worked out between human loss and the gain provided by test-tube babies.

Because human life is sacred, those who advocate the new technology have the responsibility to prove the morality of their work, rather than assuming it is all right unless someone shows otherwise. If we blindly accept whatever the genies of the laboratory make available to us, we may find forbidden fruit within the bountiful harvest and forever regret taking the first bite.

Two Into One

Along with the value of human life, the Bible supplies a second principle by which to judge the new wonders of science: the sacredness of the marriage union, especially as it relates to parenthood. According to both the Old and New Testaments, marriage and parenthood are a single, God-ordained institution. Attempts to separate sex or having children from marriage form a massive attack on the divine plan.

The Old Testament writers saw marriage as a holy symbol of the relationship between God and His people. For example, the prophet Hosea portrayed Israel as God's unfaithful wife. Jeremiah developed a similar theme, using marriage to represent the divine Covenant, and Isaiah was moved to deliver the Word of the Lord "like a wife forsaken and grieved in spirit."[16]

In the New Testament, the unity of husband and wife as one flesh, first stated in Genesis, is strongly reaffirmed by Jesus. Marriage is given as an image of the new covenant, with the church being the eternal bride. Paul speaks of a divine jealousy because the church has been betrothed to Christ, and the imagery is consummated at the Marriage Supper of the Lamb described by John in the Book of Revelation.[17]

Marriage is a fundamental order of creation. It is the normal expression of the image of God on earth, and the begetting of children is its natural completion.

As one writer states it: "Parents in the act of generation truly transcend themselves, add a newness of being, express their love for each other in a physical way, and in more than just a physical way. . . . The parents are responsible for the child and not just for the body of the child. . . . [and] it is equally true that

God is responsible for the child and not just for the soul. . . . Together God and parents create a great gift of human life."[18]

Sex outside marriage has always been a threat to its divine unity, but the present trend is to separate the two completely. Sex is thought of as an experience distinct from marriage or having children, and the arrival of new birth technologies will accelerate this radical shift.

Far from increasing our freedom as persons, though, the separation of sex and procreation from marriage turns them both into mechanical functions, and marriage becomes an empty ritual. Echoing recent thought, the U.S. Supreme Court held that marriage is just the living together by contract of separate individuals who keep their independent rights, even over their children. On this basis, the Court decided that a husband has no right to block an abortion by his wife, whether or not the child is his.[19]

The fragmentation of the marriage covenant in secular society is almost complete.

Boldly countering this disintegration, Christians know that marriage embraces the wholeness of two people—body and spirit as well as mind. All are unified through the Spirit of God, who gives the relationship meaning. Children are truly *pro*created, that is, created "on behalf of" God himself.

This is the biblical standard against which to measure the new science. It opposes artificial insemination by donor (AID), wombs-for-rent and any other procedure that would destroy the unity of father, mother and child by involving an outsider. The alternative may well be childlessness, but how does that compare to overthrowing the whole purpose of marriage? Surely, God will bless in countless other ways a couple who uphold His commands and refuse to violate the covenant of their marriage in order to have children.

In the end, the discussion comes down to a clash of beliefs, a confrontation facing each individual and couple. Nearly all of us want children, and everyone desires them to be healthy. The crucial decision for this generation is how far we are willing to go in order to reach these goals.

12

Redesigned Man

What is human life worth?

Using a new method of calculation, the Federal government can now determine our value to the nearest dollar. As of late 1978, each of us was worth an average of $287,175 to society.[1]

If, as an unborn baby, your value were weighed on this scale of cost against social benefit, would you measure up? If you had a serious genetic defect, the answer would probably be no.

According to one expert who has studied this frightening system, "The whole process has only one point: to enable us better to accumulate . . . the number of high-quality lives in the society. . . . [In this] world of limited resources, achievement of this goal has to mean investing less in lives that have fewer chances of becoming high-quality."[2]

The rush to quality can be seen first in the growing willingness to sacrifice the unborn and elderly because they interfere with our life style. As life's happiness replaces its sacredness, human values become tentative. "What is best" soon turns into "What is best for *me*," no matter what high-sounding reasoning is supplied.

Observes ethicist Bernard Ramm, "A defective child is no longer looked upon as a lesson through which we learn of God's mercy and patience with sinners, but as a terrible burden on one's time, a severe limitation of one's social life, and a heavy strain on the family's financial resources. . . . The most important factor in the minds of couples [receiving genetic counseling] . . . is not the statistical possibilities, but the sheer bother of having a defective child."[3]

Like the government cost-benefit program, the idea is spreading that human beings should be valued according to their contribution to our lives and the needs of society, rather than any inborn worth. Expert after expert insists that genetically diseased children should not be allowed to be born because they are more of a burden than a blessing, a net debit when these amateur deities total the ledger of life.

For example, in announcing its new policy of neglecting infants with serious genetic defects until they weaken and die, a Yale-New Haven Hospital spokesman explained that to have a life worth living a baby must be "lovable." Another expert writes that "no child should be admitted into the society of the living who suffers any physical or mental defect that would prevent marriage or would make others tolerate his company only from a sense of mercy."[4]

Fueling this attitude is the belief that our resources and future possibilities are steadily contracting. Many are more interested in protecting what they have than in sharing it, and when the clamor of the needy and the diseased begins to overwhelm us, society is ready to write them off altogether.

As indicated by the popularity of abortion on demand,

many couples today are willing to sacrifice their un-born baby for the sake of convenience. Could the child choose, they claim, he or she would prefer not to exist rather than live a handicapped life.

To this, John Fletcher of the Institute of Society, Ethics and the Life Sciences replies, "Whenever a strong group argues on behalf of a weaker group that their removal would be better than their survival, we should not be impressed."[5]

Fleeting feelings of personal happiness are being elevated to become life's highest goal, causing us to slowly surrender our humanity. People soon become, in the words of the great Jewish theologian Abraham Heschel, "a machine into which we put what we call food and produce what we call thought." Naturally, if the machine is not up to factory standards, it will be recalled, and if better methods of production such as cloning or test-tube assembly lines can be devised, society will gladly make use of them.

With its emphasis on specialists, modern medicine risks being caught up in this same outlook, treating patients almost as machines in the shop for repair. "American medicine today, by and large, . . . is ster-ile and not considerate of the deeper aspects of the life of the patient," comments Florida surgeon William Standish Reed. "The development of detached profes-sional impersonality has become the new demeanor of many physicians."[6] How then can we expect the medi-cal profession to have moral doubts about using the new biotechniques when they come along?

According to psychiarist Erich Fromm, man increas-ingly regards himself as a commodity in the market-place, living under the harsh judgment of such words as "useful" and "successful."[7] Our concept of what human life means has a powerful effect on how we treat people, and when value as the image of God is exchanged for worth according to what someone can *do*, we are entering the danger zone that gave rise to the death camps.

Theologian Helmut Thielicke, speaking to capacity crowds amid the rubble of postwar Germany, emphasized this point. "Once a man ceases to recognize the infinite value of the human soul," he said, "then all he can recognize is that man is something to be used. But then he will also have to go further and recognize that some men can no longer be utilized, and he arrives at the concept that there are some lives that have no value at all. . . ."[8]

Loving the Unlovable

Standing solidly against this fearful ethic is the Christian affirmation that our humanness is a basic part of us, not something to be added or taken away by current social standards. The highest value in God's eyes is not contribution to society, but expression of His love on earth, *especially toward the unlovable.*

This love built hospitals when pagan societies were leaving the sick in the streets to die; crusaded against the ancient practices of abortion and infant-killing in the first centuries of the Christian era; and gave dignity to women as equals under God rather than pieces of property.

Far from allowing us to destroy the infirm and diseased in order to increase our happiness, the Bible teaches that we should *give* our lives away for the sake of others, just as Jesus did for the very people who were crucifying Him. His love gave them meaning for the first time, a significance born in the act of His dying on their behalf.[9]

Unlike the trend of modern medicine, God does not consider health and a good genetic heritage to be the fullest realizations of humanity. Says Bernard Ramm, "It has been the contention of the Christian Church that people who suffer from illness, disease and bodily defects may nonetheless reach spiritual maturity, if not sainthood."[10]

Those who work with retarded and handicapped individuals witness to their glowing spirits despite the imperfections of mind and body. Since we are on this planet to love and serve God, our genetic weaknesses are secondary to the worship and relationship which form the heart of humanity, however distorted by hereditary defects.

Endorsing this idea (while denying its spiritual basis), biologist Jean Rostand declares, "It is an honor for a society to desire the expensive luxury of sustaining life for its useless, incompetent and incurably ill members. I would almost measure a society's degree of civilization by the amount of effort and vigilance it imposes on itself out of pure respect for life."[11]

The weaker a person, the more in need of love and compassion he is, and the church belongs in the forefront of this movement of mercy. Part of love is to make every effort to ease suffering and cure disease, and for this the new biology holds out great promise. But another part is to accept imperfections we cannot treat, rather than destroying those who are afflicted.

While we do our best to reduce illness and applaud the contributions medical and genetic technology can make, we do not have the Creator's power in our hands. If, having tried everything we know, an individual comes into the world the victim of genetic disease, he or she is still entitled to dignity and consideration as an heir to eternity. Parents were never given the absolute right to a normal child, by whatever means, and unborn babies are not interchangeable; each is a unique combination of genes which form an infinitely valuable person.

Overcoming infertility is a worthy goal, but if marriage and the family—one of the last refuges where we are loved because we *are*, not for what we can do—is ruined in the process, what have we gained? Manipulating our children before birth in order to bring them up to standard might produce healthier citizens, but what about their loss of dignity and self-worth along the way?

Reaching Our Limits

We are on the verge of great accomplishments through the new biology, but secular morality is not prepared to cope with their implications. Instead of allowing technology to set the pace for us, let us recognize our limits, both in wisdom and in sovereignty over life on this planet.

Just as we smile indulgently when our children pretend to be doctors, but would be horrified if they attempted an actual operation on one of their playmates, so God must be taking a dim view of all this preparation to re-engineer man when we have only the crudest understanding of what we are doing.

For example, one of the earliest diseases scientists hope to eliminate through genetics is diabetes. But now, just before the introduction of synthetic insulin to the mass market, doctors are discovering that a hormone from the pancreas may be responsible for much of the diabetic condition, not lack of insulin.

Furthermore, it is claimed that up to half of those diagnosed as diabetic may not really have the disease anyway, because conventional testing has proved to be misleading in many cases. And in mid-1979, childhood diabetes was strongly linked to at least one form of virus, with hope that a simple vaccine might be developed against it.[12]

What if we had begun a mass screening program, combined with abortion and genetic engineering, in an effort to wipe out diabetes? How many lives would have been needlessly lost because of our ignorance? Are we not moving too fast in the direction of manipulating life without recognizing the very serious limitations on our knowledge?

Thousands today are stricken with cancer as a result of "harmless" x-ray and atomic radiation treatments thought to be therapeutic in past decades. Limbless young people bear the scars of our ignorance because their mothers took a "safe" tranquilizer, thalidomide.

From asbestos to DDT, we find that the best ideas of one era can be the curses of the next. Yet, altering the building blocks of life is infinitely more serious an intrusion into the delicate balance of human existence than these past mistakes.

The sense of limited sovereignty and submission to God are basic to a meaningful morality. Our conscience, said Pope John Paul I in his comments on the birth of the first test-tube baby, "does not have the duty of creating law, but of informing itself first on what the law of God dictates."[13] We are free to build towers, but not towers of Babel intended to crash the gates of heaven.

A Day of Decision

Today, growing numbers are questioning whether all morality really is relative and whether science has the right to unlock the Pandora's box of life without restraint. For the first time, scientists themselves are beginning to wonder whether absolute freedom in research is in the best interests of humanity.

A new field called *bioethics* has appeared, with a few theologians and ethicists joining doctors on the hospital front lines to help both physicians and patients understand the delicate issues touched by modern medicine. Until recently, *Newsweek* magazine noted, "many doctors were too enthralled by the scientific and technological achievements of their profession to worry greatly about ethical implications. Many physicians, somewhat arrogantly, assumed that they were the sole decision makers in the care of their patients. But the consumer movement and such dramatic events as the Karen Ann Quinlan case . . . have forced the doctors to take a less imperious stance."[14]

The progress of bioethics in the medical profession is slow, but a start has been made in the right direction. Doctors are re-emphasizing the human factor, learning to possess skills without being possessed by them.

The public, too, is taking a hard look at the promised wonders of science, including the new biology. Only half the respondents in a recent *McCall's* magazine poll said they would use test-tube fertilization. Fewer than half would abort a genetically diseased child whose treatment would cause severe financial and emotional hardship, and less than twenty percent would consent to artificial insemination by a donor (AID).[15] The happiness-at-any-price ethic has not yet entrapped the whole population, although its influence is steadily growing.

We are now at the point of decision that will determine the future of human life. The moral questions raised by biotechnology are already upon us—primarily in the forms of AID and amniocentesis/abortion—and our decisions here will have far-reaching implications.

With the growing use of artificial insemination by donor, we face an attack on the fundamental meaning of marriage. To accept AID, which involves only one partner in procreation, is to give a green light to the other birth methods that split the marriage bond, including cloning, donor wombs and embryo transplants.

And the spread of amniocentesis followed by the abortion of "defectives" takes us well down the road to loss of human worth. Believing an unborn child to be foreign tissue is the first step toward manipulation in the womb and experiments aimed at breeding a new type of human being.

If any of these new procedures are to be halted or limited on moral grounds, now is the time to take a stand based on God's Word. Christians dare not be left behind in yet another moral battle.

The Bible does not resist change; in fact, its views of history and man's role were the motivating forces of modern science. But we cannot blindly accept change, either, because larger principles are involved.

"It is the fatality and the blessing of our time that its human questions always face man with the ultimate," observed theologian Helmut Thielicke. "It is always a question of rebellion or obedience. What it involves is the decision between the dogma that man can devise and manipulate and do all things, and the willingness to 'accept.' "[16]

Genetic research will—and should—continue, and we can expect it to bring exciting new benefits. Cancer may be cured, birth defects treated, infertile couples blessed with the joys of parenthood. However, rather than rush ahead and leave future generations to sweep up the rubble of our mistakes, we need to go slowly and risk holding progress back a few years. To paraphrase a well-known Bible verse, "What does it profit mankind to redesign the race if we lose our humanity in the process?"

Like most of the moral decisions that face citizens of today's wide-open society, the new biology ultimately comes down to a personal choice. Doubtless, some will happily use any technology that comes along, totally ignoring the moral implications. Others will shun them all for fear of anything radically new. Each of us ought to prayerfully decide where we fit between these two extremes, and our church leaders have a responsibility to help us understand the new developments and their promises of freedom in light of biblical truth.

The New Man

Is the New Man really coming? Many are working diligently to hasten that day, and the prospect of self-guided evolution through genetics has given a new thrust to an old dream. The goal is to recover a perfection lost so long ago in the spiritual dimension through a re-engineering of man in the physical. Instead, the manipulations would at best enslave future generations to our own ideas and desires, at worst revive the horrors of Nazism.

But God has provided another way to perfection, a path of real freedom stretching beyond both genes and environment. When God said, "Let us make man in our own image," the image He had in mind was Jesus Christ, the Model Man who was the perfection of God in human form.[17]

With a brain the size of ours, He had all wisdom and knowledge, exceeding the limits of our worldly dimension. Having a body similar to our own, He was able to step through walls and instantly speed to the depths of space. Possessing a will like ours, He submitted it in obedience to His Father, showing us the proper way to reach our potential as human beings.[18]

Health, intelligence and heredity were secondary with Him. Purity, love and faith are the qualities God considers important, and nothing science can produce has any influence on these essentials of full humanity.

We live on a planet that once knew perfection, and man lived in the unspoiled image of God. Because of the same rebellious spirit that today drives attempts to redesign humanity, we lost the beauty and meaning that belonged to our race. As a result, we are surrounded by mental and physical disease, crippled children and infirm old age.

It is only natural that we try to eliminate these results of the evil that plagues our world, and biotechnology can be an asset to our efforts. But while Adam was given command over the plant and animal kingdoms by God, nowhere was he granted authority over his own species.[19] That is the special province of our Creator. We should never forget that among the first recorded words of Satan are, "You shall be as gods."[20]

The New Man *is* coming, in the person of the risen Christ, and the greatest wonder in the universe is that we will be transformed to become like Him—if we submit to His authority and lordship during our life on earth.[21] Rather than trampling the divine will in a futile effort to become as gods, our calling is to accept His sovereignty and thus fulfill the infinite vision He has for us.

God's promise to us goes far beyond the halfway measure of improving man physically while the heart and soul continue to rot. He pledges to redesign us completely, inside and out, and we don't need to wait for heaven before the process begins. "We are being changed into His image from glory to glory," Paul affirmed, as we walk with Jesus through His Spirit.[22]

Growing ever closer to the ideal intended by God with the first man and woman, we will suddenly be transformed at the appearance of Christ, whisked away to spend eternity in His presence. There, we will be forever clothed in radiant splendor, perfectly redesigned by a sure hand and living at last up to our potential as God's beloved children.

Strolling together on streets of dazzling gold along a crystal-sparkling river, we will radiate the joy of complete fulfillment. Senses will be heightened—exquisite fragrances and the blending of harmonious sounds will add to the beautiful sights, all bathed in a gentle light of holiness. Understanding will be instantaneous, weariness, illness and sorrow unknown in our shimmering new bodies.[23]

The simplest description of redesigned man as God intends us turns the most exciting possibilities of modern genetics to a lifeless gray. All this is ours if we will accept God's plan for rebuilding humanity instead of creating our own.

"Man," said Helmut Thielicke, "is the risk of God. An animal cannot fail to fulfill its destiny, but . . . you and I can. We can play the wrong role. And at the Last Judgment there may be written on the margin of our life in red ink: 'You missed the point.'"[24]

Our mission is to lovingly change what we can in human life and humbly accept what we must. By prayerfully considering how God's Word applies to unprecedented moral situations, we not only can discover God's will in the age of biotechnology, but—most important—learn to know Him better.

And that is the "point," the very reason He said, "Let us make man."

Source Notes

Chapter 1

[1]George Gordon, "The World's First Test-Tube Baby," *Star*, August 1, 1978, p. 16.

[2]"Test-Tube Baby: It's a Girl," *TIME*, August 7, 1978, p. 67.

[3]Lesley Brown, "I'd Like Another Little Miracle," *Vancouver Sun*, August 19, 1978, p. A13.

[4]"The First Test-Tube Baby," *TIME*, July 31, 1978, p. 59.

[5]Ibid., p. 62.

[6]U.S. Department of Health, Education and Welfare Ethics Advisory Board, *Hearing*, San Francisco, California, November 14, 1978, p. 149.

[7]Peter Gwynne, et. al., "All About That Baby," *Newsweek*, August 7, 1978, p. 71.

[8]George F. Will, "Our New Baby Technology May Grow Into a Monster," *Los Angeles Times*, July 30, 1978, Part VI, p.5.

[9]Gerald George, "Creating Life in the Lab," *Science Digest*, January 1974, p. 33.

[10]Ted Howard and Jeremy Rifkin, *Who Should Play God?* (New York: Dell Books, 1977), p. 174.

[11]Edward Grossman, "The Obsolescent Mother," *Atlantic Monthly*, May 1971, pp. 48-49.

[12]Robert L. Sinsheimer, "The Prospect of Designed Genetic Change," in Adela S. Baer, ed., *Heredity and Society* (New York: Macmillan Co., 1973), p. 356.

[13]Howard and Rifkin, *Play God*, p. 81.

[14]Charles Frankel, "The Specter of Eugenics," *Commentary*, March 1974, p. 28.

[15]Rev. 13:16-18.

[16]Caryl Rivers, "Genetic Engineers," *Ms.*, June 1976, p. 120.

[17]Howard and Rifkin, *Play God*, p . 88.

[18]Subcommittee on Health and the Environment of the House Committee on Interstate and Foreign Commerce, *Hearings*, March 15, 1977, p. 79.

[19]Grossman, "Obsolescent Mother," p. 47.

[20]Frankel, "Specter," p. 33.

[21]Grossman, "Obsolescent Mother," p. 43.

[22]Rivers, "Engineers," p. 120.

[23]Subcommittee on Science, Technology and Space of the Senate Committee on Commerce, Science and Transportation, *Hearings*, November 10, 1977, p. 338.

[24]June Goodfield, *Playing God* (New York: Random House, 1977), p. 134.

[25]Joseph Fletcher, *The Ethics of Genetic Control* (Garden City, New York: Doubleday Anchor Books, 1974), p. 112.

Chapter 2

[1]Vance Packard, *The People Shapers* (New York: Bantam Books, 1977), p. 266.

[2]*TIME*, November 25, 1974, p. 67.

[3]Packard, *People Shapers*, p. 265.

[4]David Rorvik, *Brave New Baby* (New York: Pocket Books, 1971), p. 43.

[5]Packard, *People Shapers*, pp. 269-70.

[6]Robert T. Francoeur, *Utopian Motherhood* (Garden City, New York: Doubleday, 1970), p. 58.

[7]*Los Angeles Times*, May 17, 1971, Part IV, p. 1.

[8]Edward Grossman, "The Obsolescent Mother," *Atlantic Monthly*, May 1971, p. 48.

[9]Rorvik, *Brave New Baby*, p. 99.

[10]Gerald Leach, *The Biocrats* (New York: McGraw-Hill, 1970), p.153

[11]Albert Rosenfeld, *The Second Genesis* (Englewood Cliffs, N.J.: Prentice-Hall, 1969), p. 117.

Chapter 3

[1]Pat K. Lynch, "Women: The Next Endangered Species?"

Source Notes

Madamoiselle, May 1977, p. 32.

[2]*Los Angeles Times*, February 16, 1979, Part IV, p. 22.

[3]David Rorvik, *Brave New Baby* (New York: Pocket Books, 1971), p.64.

[4]Ibid., p. 67.

[5]Vance Packard, *The People Shapers* (New York: Bantam Books, 1977), pp. 288-89. Charles F. Westoff and Ronald R. Rindfuss, "Sex Preselection in the United States: Some Implications," *Science*, May 10, 1974, p. 636.

[6]Amitai Etzioni, "Sex Control, Science and Society," in Adela S. Baer, ed., *Heredity and Society* (New York: Macmillan Co., 1973), pp. 238-39.

[7]Lynch, "Women," p. 34.

[8]Marc Lappe′, "Choosing the Sex of Our Children," *The Hastings Center Report*, Institute of Society, Ethics and the Life Sciences, February 1974, p. 3.

[9]*Los Angeles Times*, October 13, 1978, Part I, p. 15.

[10]Gerald Leach, *The Biocrats* (New York: McGraw-Hill, 1970), pp. 74-74.

[11]Philip Reilly, *Genetics, Law and Social Policy* (Cambridge, Mass.: Harvard University Press, 1977), pp. 190-91.

[12]"Who Is Daddy?" *Weekend*, NBC-TV, aired December 2, 1978.

[13]*Newsweek*, February 12, 1979, p. 61.

[14]Ibid.

[15]Ted Howard and Jeremy Rifkin, *Who Should Play God?* (New York: Dell Books, 1977), p. 27.

[16]Ibid., pp. 95-96; Packard, *People Shapers*, pp. 232-33.

[17]Packard, *People Shapers*, p. 230.

[18]Richard M. Restak, *Pre-Meditated Man* (New York: Viking Press, 1975), pp. 69-70.

[19]*Los Angeles Times*, December 27, 1978, Part IV, p. 4.

[20]Ibid.

[21]*TIME*, February 25, 1966, p. 48.

[22]Joseph Fletcher, *The Ethics of Genetic Control* (Garden City, New York: Doubleday Anchor Books, 1974), pp. 15, 87.

[23]Gen. 16:1-4, 21:9-14.

[24]J.A. Thompson, *The Bible and Archaeology* (Grand Rapids, Mich.: Wm. B. Eerdmans, 1972), pp. 28-30.

[25]Gen. 2:24; Eph. 5:31.

[26]Matt. 19:9; 1 Cor. 6:16.

LET US MAKE MAN

Chapter 4

[1]Martin Ebon, *The Cloning of Man* (New York: Signet Books, 1978), p. 114.

[2]Aldous Huxley, *Brave New World* (New York: Harper and Row, 1946), p. 3.

[3]The four ways genes can pair off are as follows: DOMINANT-DOMINANT; DOMINANT-recessive; recessive-DOMINANT; recessive-recessive. Only in the latter case would the recessive trait appear in the offspring, being masked by the dominant trait in the others.

[4]Allen R. Utke, *Bio-Babel* (Atlanta: John Knox Press, 1978), p. 107.

Chapter 5

[1]Anthony Smith, *The Human Pedigree* (Philadelphia: J.B. Lippincott, 1975), pp. 19-20.

[2]The chances of inheriting genetic disease can be calculated in exactly the same way Gregor Mendel determined the traits of his garden peas. It should be kept in mind that the odds are the same for *each* child, no matter how many children have already been born. Thus, if the chances of disease are fifty-fifty, having one diseased child does *not* guarantee the second will be healthy, and vice versa.

[3]*Newsweek*, January 8, 1979, p. 41.

[4]Jean Ashton, "Amniocentesis: Safe But Still Ambiguous," *Hastings Center Report 6* (February 1976), p. 6.

[5]Paul Ramsey, "Screening: An Ethicist's View," in Bruce Hilton, et. al., eds., *Ethical Issues in Human Genetics* (New York: Plenum Press, 1973), p. 157.

[6]Aubrey Milunksy, et. al., "Prenatal Genetic Diagnosis," *New England Journal of Medicine*, December 31, 1970, p. 1504. Italics added.

[7]David Hendin and Joan Marks, *The Genetic Connection* (New York: William Morrow, 1978), p. 208.

[8]Richard M. Restak, *Pre-Meditated Man* (New York: Viking Press, 1975), p. 83.

Chapter 6

[1]Edwin Kiester, Jr., "Healing Babies Before They're Born,"

Source Notes

Family Health, October 1977, pp. 26-28.

[2]Horace Freeland Judson, "Fearful of Science," *Harper's*, March 1975, p. 41.

[3]Ruth Halcomb, "Winning the War Against Birth Defects," *Parents' Magazine*, May 1977, p. 54.

[4]Rick Gore, "The Awesome Worlds Within a Cell," *National Geographic*, September 1976, p. 386.

[5]Leroy Augenstein, *Come, Let Us Play God* (New York: Harper and Row, 1969), pp. 103-104.

[6]Lucy Eisenberg, "Genetics and the Survival of the Unfit," *Harper's*, February 1966, p. 53.

[7]Ellen Goodman, "Eliminating Defects Before They Are Born," *Los Angeles Times*, June 24, 1979, Part V, p. 5.

[8]Judson, "Fearful," p. 40.

[9]Marc Lappe', et, al., "Ethical and Social Issues in Screening for Genetic Disease," *New England Journal of Medicine*, May 25, 1972, p. 1131.

[10]Laurence E. Karp, M.D., *Genetic Engineering: Threat or Promise?* (Chicago: Nelson-Hall, 1976), p. 81.

[11]Vance Packard, *The People Shapers* (New York: Bantam Books, 1977), p. 318.

[12]Paul Ramsey, *Fabricated Man* (New Haven: Yale University Press, 1970), p. 35.

[13]Daniel Callahan, "The Meaning and Significance of Genetic Disease," in Bruce Hilton, et. al., eds., *Ethical Issues in Human Genetics* (New York: Plenum Press, 1973), p. 88.

[14]Leon Kass, "Implications of Prenatal Diagnosis for the Human Right to Life," in Hilton, *Issues*, p. 189.

[15]Richard M. Restak, *Pre-Meditated Man* (New York: Viking Press, 1975), p. 89.

[16]Paul Ramsey, "Screening: An Ethicist's View," in Hilton, *Issues*, pp. 158-59.

[17]Gerald Leach, *The Biocrats* (New York: McGraw-Hill, 1970), p. 177.

[18]Isaiah 45:9-11.

Chapter 7

[1]Marc Hillel and Clarissa Henry, *Of Pure Blood*, trans. by Eric

Mossbacher (New York: McGraw-Hill, 1976).
[2]Ibid., pp. 16, 244.
[3]Ibid., pp. 213-14.
[4]Anthony Smith, *The Human Pedigree* (Philadelphia: J.B. Lippincott, 1975), pp. 75-76.
[5]Ibid., p. 82.
[6]Kenneth M. Ludmerer, *Genetics and American Society* (Baltimore: Johns Hopkins Press, 1972), p. 11.
[7]Ted Howard and Jeremy Rifkin, *Who Should Play God?* (New York: Dell Books, 1977), p. 47.
[8]Ludmerer, *Genetics*, pp. 28-29.
[9]Howard and Rifkin, *Play God*, p. 58.
[10]Ibid., p. 55.
[11]Ibid., p. 69.
[12]Ibid., p. 57.
[13]Allan Chase, "The Great Pellagra Cover-Up," *Psychology Today*, February 1975, pp. 83-86.
[14]D.S. Halacy, Jr., *Genetic Revolution* (New York: Harper and Row, 1974), p. 93.
[15]Curt Stern, "Genes and People," in Adela S. Baer, ed., *Heredity and Society* (New York: Macmillan Co., 1973), p. 316.

Chapter 8

[1]Joseph Fletcher, *The Ethics of Genetic Control* (Garden City, New York: Doubleday Anchor Books, 1974), p. 199.
[2]Frederick Ausubel, Jon Beckwith and Kaaren Janssen, "The Politics of Genetic Engineering," *Psychology Today*, June 1974, p. 34.
[3]Ibid., p. 40.
[4]Anthony Smith, *The Human Pedigree* (Philadelphia: J.B. Lippincott, 1975), p. 148.
[5]Ibid.
[6]Bentley Glass, "Science: Endless Horizon or Golden Age?" *Science*, January 8, 1971, p. 28.
[7]Smith, *Pedigree*, p. 247.
[8]Amitai Etzioni, *Genetic Fix* (New York: Macmillan, 1973), p. 108.
[9]Philip Reilly, *Genetics, Law and Social Policy* (Cambridge: Harvard University Press, 1977), p. 135. Italics added.

Source Notes

[10]Robert T. Francoeur, *Utopian Motherhood* (Garden City, New York: Doubleday, 1970), p. 218.

[11]Charles Frankel, "The Specter of Eugenics," *Commentary*, March 1974, p. 25.

[12]David Rorvik, *Brave New Baby* (New York: Pocket Books, 1971), pp. 113-15.

[13]Albert Rosenfeld, *The Second Genesis* (Englewood Cliffs, New Jersey: Prentice-Hall, 1969), pp. 142-43.

[14]Rorvik, *Baby*, p. 117.

[15]Arnold B. Barach, *1975 and the Changes to Come* (New York: Harper and Row, 1962).

[16]Paul Ramsey, *Fabricated Man* (New Haven: Yale University Press, 1970), p. 96.

[17]Marc Lappe', "Moral Obligations and Fallacies of 'Genetic Control,' " in Thomas A. Shannon, ed., *Bioethics* (New York: Paulist Press, 1976), p. 358.

[18]Nicholas Wade, *The Ultimate Experiment* (New York: Walker and Co., 1977), p. 150. Italics added.

Chapter 9

[1]*Modern People*, April 18, 1978, p. 1.

[2]Martin Ebon, *The Cloning of Man* (New York: Signet Books, 1978), pp. 80-81, 84.

[3]*TIME*, July 24, 1978, p. 47.

[4]David Rorvik, *In His Image* (New York: Pocket Books, 1978), pp. 67-68.

[5]Ted Howard and Jeremy Rifkin, *Who Should Play God?* (New York: Dell Books 1977), p. 123.

[6]D.S. Halacy, Jr., *Genetic Revolution* (New York: Harper and Row, 1974), p. 163.

[7]Ibid., p. 160.

[8]Robert T. Francoeur, *Utopian Motherhood* (Garden City, New York: Doubleday, 1970), p. 138.

[9]Rorvik, *Image*, p. 41.

[10]*TIME*, April 19, 1971, p. 51.

[11]Horace Judson, "Fearful of Science," *Harper's*, March 1975, p. 39.

[12]Gerald Leach, *The Biocrats* (New York: McGraw-Hill, 1970), p. 97.

13Paul Ramsey, *Fabricated Man* (New Haven: Yale University Press, 1970), pp. 71-72.

14Jer. 1:5; Isa. 49:1-5; Psalms 139:14-16.

15Albert Rosenfeld, *The Second Genesis* (Englewood Cliffs, New Jersey: Prentice-Hall, 1969), p. 130.

16*Newsweek*, April 17, 1978, p. 16.

17*Newsweek*, February 12, 1979, p. 99.

18James D. Watson, "Moving Toward Clonal Man," *Atlantic Monthly*, May 1971, p. 53.

Chapter 10

1"A Unique Plan for Soviet Molecular Biology," *New Scientist*, January 8, 1976, p. 53.

2Victor Zorza, "Spectre of a Genetic 'Arms Race,' " *Guardian Weekly*, December 13, 1969, p. 6.

3Subcommittee on Science, Technology and Space of the Senate Committee on Commerce, Science and Transportation, *Hearings on Regulation of Recombinant DNA Research*, November 2, 8 and 10, 1977, pp. 423-24.

4Ted Howard and Jeremy Rifkin, *Who Should Play God?* (New York: Dell Books, 1977), p. 41.

5June Goodfield, *Playing God* (New York: Random House, 1977), p. 58.

6Maxine F. Singer, "The Recombinant DNA Debate," *Science*, April 8, 1977, p. 3.

7*Newsweek*, January 12, 1976, p. 52.

8Subcommittee on Science, Research and Technology of the House Committee on Science and Technology, *Report on Science Policy Implications of DNA Recombinant Molecule Research*, March 1978, pp. 3-4.

9Ibid., p. 10.

10*Federal Register*, vol. 44, no. 118 (June 18, 1979), p. 35057.

11Robin Marantz Henig, "Go Forth and Multiply," *BioScience*, May 1979, pp. 321-23.

12Subcommittee on Health and the Environment of the House Committee on Interstate and Foreign Commerce, *Hearings on Recombinant DNA Regulation Act of 1977*, March 15-17, 1977, p. 91.

Source Notes

Chapter 11

[1]Robert Meyers, "Like Normal People," *Family Circle*, August 7, 1978, pp. 6, 8 and 14.

[2]Gen. 11:6.

[3]Albert Rosenfeld, "When Man Becomes as God," *Saturday Review*, December 10, 1977, p. 19.

[4]D. Gareth Jones, "Making New Man," *Journal of the American Scientific Affiliation*, December 1974, p. 145.

[5]Gen. 1:27.

[6]Gen. 9:6; James 3:9.

[7]G.C. Berkouwer, *Man: The Image of God* (Grand Rapids, Mich.: Wm. B. Eerdmans, 1962), pp. 75, 92 and 201.

[8]Jean Rostand, *Humanly Possible*, trans. Lowell Bair (New York: Saturday Review Press, 1973), p. 89.

[9]Edward O. Wilson, *On Human Nature* (Cambridge: Harvard University Press, 1978), pp. 5-6.

[10]*Newsweek*, October 16, 1978, p. 118; Richard Dawkins, *The Selfish Gene* (New York: Oxford University Press, 1976), p. 21.

[11]C.S. Lewis, *God in the Dock*, ed. Walter Hooper (Grand Rapids, Mich.: Wm. B. Eerdmans, 1970), p. 137. An illustration of just how irrational the basis of evolution is comes from the renowned geneticist Jacques Monod. In his bestselling 1971 book, *Chance and Necessity*, he maintains that the origin of life now looks like it was a unique event, occurring just once by chance, with a probability of nearly zero. The irony is that this same point—the uniqueness and improbability of the events—has been used for centuries by opponents of Christianity to "disprove" the possibility of divine miracles, including that of Creation!

[12]Dwight J. Ingle, *Who Should Have Children?* (Indianapolis: Bobbs-Merrill, 1973), p. 111; Joseph Fletcher, "Indicators of Humanhood," *The Hastings Center Report*, vol. 2, no. 5 (November 1972); Joseph Fletcher, *The Ethics of Genetic Control* (Garden City, New York: Doubleday Anchor Books, 1974), p. 171.

[13]HEW Ethics Advisory Board, *Hearings*, Bethesda, Maryland, September 15, 1978, p. 255.

[14]C. Everett Koop, "Medical Ethics and the Stewardship of Life," *Christianity Today*, December 15, 1978, pp. 10-11.

[15]Fletcher, *Genetic Control*, pp. 142, 185.

[16]Jer. 31:31f; Isa. 54:6.

[17]Matt. 19:4-6; 2 Cor. 11:2-4; Eph. 5:21-23; Rev. 19:6-16.

[18]John F. O'Grady, *Christian Anthropology* (New York: Paulist Press, 1976), p. 106.

[19]Paul Ramsey, *Ethics at the Edge of Life* (New Haven: Yale University Press, 1978), pp. 9-10.

Chapter 12

[1]Fred Hapgood, "Risk-Benefit Analysis," *Atlantic Monthly*, January 1979, p. 35.

[2]Ibid.

[3]Bernard Ramm, "Ethical Evaluation of Bio-genetics," *Journal of the American Scientific Affiliation*, December 1974, p. 138.

[4]Paul Ramsey, *Ethics at the Edge of Life* (New Haven: Yale University Press, 1978), pp. 205-6.

[5]John Fletcher, "Parents in Genetic Counseling," in Bruce Hilton, et. al., eds., *Ethical Issues in Human Genetics* (New York: Plenum Press, 1973), p. 323.

[6]William Standish Reed, M.D., *Surgery of the Soul* (Old Tappan, New Jersey: Fleming H. Revell Spire Books, 1969), pp. 122, 127.

[7]Charles R. Stinette, *Faith, Freedom and Selfhood* (Greenwich, Connecticut: Seabury Press, 1959), p. 31.

[8]Helmut Thielicke, *Nihilism*, trans. John W. Doberstein (New York: Schocken Books, 1969), pp. 84, 114-15.

[9]Matt. 16:25; Phil. 2:3-4; 1 John 4:10.

[10]Ramm, "Ethical Evaluation," p. 142.

[11]Jean Rostand, *Humanly Possible*, trans. Lowell Bair (New York: Saturday Review Press, 1973), pp. 89-90.

[12]National Public Radio *Report*, May 24, 1979; *Star*, September 26, 1978, p. 7.

[13]*TIME*, September 4, 1978, p. 66.

[14]*Newsweek*, January 8, 1979, p. 67.

[15]Susan Jacoby, "Our Unborn Children: The Disturbing New Choices," *McCall's*, March 1979, pp. 130-31.

[16]Helmut Thielicke, *The Ethics of Sex*, trans. John W. Doberstein (New York: Harper and Row, 1964), p. 268.

[17]2 Cor. 3:18, 4:4; Col. 1:15; Heb. 1:3.

[18]Matt. 7:29, 13:54, 26:39; John 20:19; 1 Cor. 1:30; Phil. 2:8; Col. 1:16.

Source Notes

[19]Gen. 1:28-29.
[20]Gen. 3:5.
[21]Rom. 8:19-24; 1 Cor. 15:51-53; 1 John 3:2-3.
[22]2 Cor. 3:18.
[23]1 Cor. 15:47-50; Rev. 21:1-4, 10-27, 22:1-5.
[24]Helmut Thielicke, *How the World Began*, trans. John W. Doberstein (Philadelphia: Fortress Press, 1961), p. 76.